DIGITAL DILEMMAS

Ethical Issues for
Online Media Professionals

DIGITAL DILEMMAS

Ethical Issues for
Online Media Professionals

Robert I. Berkman
Christopher A. Shumway

Media and Technology Series
Alan B. Albarran, Series Editor

Iowa State Press
A Blackwell Publishing Company

Robert I. Berkman is on the faculty of the Media Studies M.A. Program at New School University, New York, NY. He is the author of several books on online searching and research and the founder and editor of *The Information Advisor,* a monthly journal for professional researchers.

Christopher A. Shumway was a broadcast journalist for 15 years, serving as a photographer, news reporter, anchor, and meteorologist at local TV stations in Alabama, Georgia, Texas, and Ohio. During that time he won many awards for his work, including a regional Emmy nomination. Chris has a Master's degree in Media Studies from New School University, New York, NY.

Iowa State Press
2121 State Avenue, Ames, Iowa 50014

Orders: 1–800–862–6657
Office: 1–515–292–0140
Fax: 1–515–292–3348
Web site: www.iowastatepress.com

First edition, 2003

Library of Congress Cataloging-in-Publication Data

Berkman, Robert I.
    Digital dilemmas: ethical issues for online media professionals/
Robert I. Berkman, Christopher A. Shumway.
        p. cm—(Media and technology series)
Includes bibliographical references and index.
    ISBN 0-8138-0236-9 (alk. paper)
    1. Journalistic ethics. 2. Electronic journals. I. Shumway,
Christopher A. II. Title. III. Series.
    PN4756.B47 2003
    175—dc21   2003009335

The last digit is the print number: 9 8 7 6 5 4 3 2 1

To Cecelia

Contents

Part III
Journalistic Ethical Dilemmas
for Online Media Professionals   219

Foreword

In *Digital Dilemmas: Ethical Issues for Online Media Professionals*, authors Robert I. Berkman and Christopher A. Shumway provide an important contribution to our understanding of the challenges facing journalists in an exploding era of technological capabilities. The authors raise a number of critical questions, one of the most important being, Who should be called a journalist?

Focusing throughout the book on ethical codes of conduct for online media professionals, Berkman and Shumway successfully guide the reader through many different topics to illustrate the complicated and controversial intersection of technology, ethics and regulation amid a host of social issues such as privacy, free speech and the protection of intellectual property. Using real world examples as well as hypothetical cases, the authors challenge the reader to carefully analyze and evaluate ethical decision making in an online journalism world.

But *Digital Dilemmas: Ethical Issues for Online Media Professionals* does not just deal with a theoretical discussion of ethics for online media professionals. The authors also detail the practical pressures of accurate storytelling in a real-time delivery system, the need for better online research, and the continuing discussion of the functions of advertising versus editorial content. The book also addresses the growing consolidation of the Internet and the mass media and their impact on democracy and democratic values.

Digital Dilemmas: Ethical Issues for Online Media Professionals will be of interest to many audiences, including undergraduate and graduate students, journalism faculty and professional journalists. As a member of the Iowa State Press Series on Media and Technology, *Digital Dilemmas* is a welcome addition to the literature and a work that readers should find stimulating, thought provoking and compelling.

Alan B. Albarran, Ph.D.
The University of North Texas
Series Editor, Media and Technology

Preface

One day in the spring of 1998 I sat at a tiny Greenwich Village café having lunch with the director of the New School University's media studies graduate program. He wanted to meet and discuss his interest in developing a course that would explore some of the contentious social issues that were being caused by the growth of the Internet. We discussed privacy, speech and intellectual property—very big issues then, and still big issues now.

As we talked, it became clear that the media professionals that worked on the Internet—the "new media" as it was popularly called during the 1990s—were facing tricky ethical issues related to these areas. And there were little legal or even clear professional guidelines to help in making a good, ethical decision when confronting these novel situations. Most of the discussion for those studying the media and the Internet at that point revolved around practical matters such as how to conduct e-commerce, effective Web design and online research strategies. Others focused on the technological possibilities of the medium that lay in the future. But there wasn't a course examining the big picture—the larger social issues and matters that had clear ethical implications for all of us.

Based on that lunchtime conversation, I created a course for the department that would surface and examine ethical issues of importance to media professionals working on the Internet. The purpose of the class was to provide an ethical framework to work through some of these difficult issues.

In developing this course I was unable to locate any single text that discussed privacy, speech and intellectual property together, specifically with an ethical perspective. So instead, I found one outstanding book on communications ethics called *Ethics in Communication* and used this as the primary text to create an ethical compass for the class to rely on. I then supplemented that book with a couple of others on journalism ethics. Finally, I added several other books that examined how privacy, speech and intellectual property were intersecting with the Internet. Together, these were used as a jumping-off point so that media students could have an ethical framework for deciding how to approach and resolve these "digital dilemmas."

But what is the real purpose of this book? Why is it important to understand the ethical issues that face just *online* media professionals?

A few years after I began my course I participated in an online discussion with fellow faculty members who were recently assigned to teach an online course. The purpose of this online discussion was to introduce faculty who were new to online teaching to those who had been teaching in this manner for some time, so that the newer teachers could ask questions and tap into some of their colleagues' experiences. I introduced myself to the other instructors and told them that I was teaching a class called "New Media Ethics." This was a course I had created a few years ago for the New School's M.A. in media studies program and had taught online for a few years.

One of the other faculty in this online discussion was a journalist, and he was curious about the nature of my course. Actually, my colleague was politely critical. Why, he wondered, was there a need for a separate course in ethics for the *new* media—why shouldn't the existing principles and agreed-upon standards of the current ethical codes be good enough to apply to those journalists working in any medium, including the Internet? He was concerned that my course implied that somehow the ethical standards for media working online were different, or worse, lower than for print or broadcast.

I keyed back a reply, stating that I was in fact in agreement with him, that the existing professional ethics shouldn't depend on the particular medium. However, I pointed out there were many new and unanticipated issues that have arisen for media and communication professionals that have been performing their job online. So it was necessary to surface, discuss and figure out how these existing ethics apply to these tricky new areas and perhaps to identify additional ethical standards where needed. This was what my New Media Ethics course was all about.

I don't know if my colleague was completely persuaded. But I am convinced that understanding the new ethical challenges of the new media and making the right decisions are vital for anyone who works in the online media.

This book covers all of these ethical matters in one place. It describes the existing ethical principles governing the media and then surfaces ethical dilemmas that media professionals face when working on the Internet in the areas of privacy, speech and intellectual property—as well as a few new areas.

Why This Book

It's no secret that the general public does not hold the media in high regard. The condition of print and broadcasting journalism in the United States has been uninspiring for some time. There's been the obvious emphasis on shallow entertainment and celebrity reporting; a lack of commitment by upper management and ownership to invest in staffing to undertake original, investigative reporting; regular occurrences of insensitive reporting and invasions of privacy; and a disturbing increase in reports of the business side of news making incursions into the traditional domain of editorial.

Now there is a new medium for journalists and mass communication professionals: the Internet. Many have expressed worry that the Internet will mean an even greater diminishment in media standards. Media professionals working online face new potential perils and opportunities for unethical activity—*electronic* invasions of people's privacy, trading accuracy for speed, further blurring the line between editorial and advertising, and other pitfalls discussed in the pages of this book.

These concerns do not have to come to pass because the future is still being written—right now—by those who work in it. The digital medium is still new, emerging and in flux. There are many more questions than there are answers. The exciting thing is that because it's all in flux, we can all have a part in shaping it, and in fact even help create what its future is going to look like. As those who study it know well, technology never develops in a vacuum. Its ultimate form is dependent on the social, cultural, political and other forces that shape it as it develops. The decisions that today's online media professionals make, and the guiding principles they choose to adopt, are some of those forces that will help shape this new medium.

So the actions you personally take now early on in the medium's development will serve to shape the way it evolves and will also form the practices of online media. I hope that this book provides you with a way to frame the ethical issues that you will face while working online, so you can be prepared to anticipate, think about and ultimately make the kinds of decisions you feel good about as you do your job in this exciting and still evolving medium.

Finally, I have greatly enjoyed working with my colleague and former New Media Ethics student Chris Shumway. Chris has been of incalculable assistance, providing his meticulous research, in-depth exploration of case studies, and substantive contributions, and has done much of the work to help bring the concepts of the class to the pages of this book.

Robert I. Berkman

Acknowledgments

My sincere thanks to those who spurred my interest in the topic of media ethics. The media ethics seminar I took in 1995 from ethics professor Deni Elliot while studying for my M.A. in journalism at the University of Montana inspired me to continue my investigations in this field. That seminar also introduced me to an invaluable text, *Groping for Media Ethics in Journalism* by Ron Smith, a work I've used as a touchstone on the topic and one that I often called upon when writing this book. I am also grateful to my other professors from the University of Montana. In particular, I'd like to thank Charles E. Hood, dean emeritus, who made my time both enjoyable and productive. Dennis Swibold and Clem Work are two other teachers that I had the privilege to get to know well and learn from, and I thank them as well.

The students in my New Media Ethics class at New School University regularly supply me with new ideas and keep me up to date. And I'm grateful to New School University's chair of the Department of Communications, Carol Wilder, for her interest in taking on the New Media Ethics course, which was the original impetus behind the book.

As I note in the preface, I relied greatly on Chris Shumway to perform research, to help me think through issues, and to put my initial concepts into final book form. Chris is an extraordinarily talented researcher, writer and thinker, and it is hard to conceive of this book coming to fruition without his assistance.

I'd also like to thank my wife Mary for reading through my early chapters and lending her sharp eyes and her always invaluable insights. And, last but not least, I'd also like to express thanks to my stepdaughter Cecelia, to whom this book is dedicated. Her admirable capacity to analyze the public sphere and the media has broadened my thinking to my own benefit and to the benefit of this book.

Robert I. Berkman

Introduction

Consider the following:

- A seasoned reporter of a major print newspaper is assigned to that newspapers' companion Web site and is asked to write a story on an increase in anorexia. During her research, she is shocked to discover an online chat site where anorexics give other adolescent girls tips on how to do a "better job" at avoiding eating and becoming even thinner. The reporter quietly joins this chat room to listen in to the discussion for her story, not identifying herself as a reporter. She uses remarks made by the participants in her story, not quoting them by name, but without asking for permission to do so.
- A young man just out of journalism school has just joined the staff of a hip new, stand-alone Web site that provides news and advice on retro style entertaining. He is writing a piece about where to find jadeite dishware and locates a site on the Web with some really sharp graphics of these dishes, and so he copies them for use in his article. In the article a store is mentioned that he visited that had a particularly impressive collection and the online reporter includes a link to that store's Web site at the end of his article.
- A smart, technical savvy college student has decided to start her own personal news site (what are called "Weblogs") on the topic of media coverage of anti-war activity. She quickly attracts a large online audience of loyal readers who like her no-holds-barred commentary, lively opinions and lists of useful new links on the Web. This particular "blogger" feels that the traditional media are an elite, unresponsive group that only tells the public what it wants

them to hear. She doesn't want to be constrained by their old rules, which include verifying and confirming data, so whenever she hears an interesting rumor she puts it up on the Web. If it's wrong, she says to herself, one of my readers will let me know about it, and I'll fix it.

What were your feelings when you read about these cases? Did you note the ethical dilemmas? In the first case, there are questions about privacy: What are the privacy rights of those who participate in an on-line discussion group? In the second, there are two issues of concern. One regards intellectual property and the other revolves around the journalistic ethic of distinguishing editorial from commercial. Was it appropriate to lift someone else's images just because they were "freely available" on the Web? Did the reporter's linking to a commercial Web site violate the traditional wall between advertising and editorial matter? In the final case, the questions raised are even larger—should this person even be called a journalist? Does she have the same obligations as those who work in the traditional media?

While these are fictional scenarios, they are in fact representative of the kinds of decisions and situations that those who work in the online media are facing and are the kinds of situations examined in this book.

It is not always clear how to best resolve these kinds of new dilemmas that arise specifically from working in the Internet medium. The laws may not be clear, and even the existing professional standards may not offer much help. As a media professional working in the new medium, you will want to be sure that when you make a difficult decision it is made from a solid grounding and with a clear understanding of the ethical ramifications.

Perhaps you are a journalist working on the Web version of a nationally known print newspaper or broadcasting network. Or you have joined a small but serious news site start-up on the Net, without backing from any larger organization. Maybe you have developed your own Weblog and serve a particular niche you feel passionate about and people rely on you for your opinion and advice. Some of you are in school, studying for your M.A. in journalism or other media studies degree with an interest in working on the Internet. In any case you are a current or future online media professional.

As an online media professional, you will need specific guidance in how to apply and interpret these basic principles to cover the new challenges from working in the digital world. And that is the purpose of this book—to help you understand what are likely to be some of the most challenging and novel dilemmas you are likely to face when working online, and what are some basic ethical principles that can serve as a foundation for thinking through and working through them. In this way, you can face these dilemmas head on and arrive at a satisfying—and ethical—decision.

Book Structure

The material in this book is organized in three major parts. In part 1 (chapters 1, 2 and 3) we lay the foundation and groundwork. Here we'll review the importance of ethics in the media, what kinds of ethical standards and codes for media professionals already exist, and whether these existing codes can be applied to those who work in the online environment. We'll also explore the historical role of how journalism is intertwined with democracy, and then discuss a very new and contentious question: In a time when anyone can disseminate news and opinions on the Web to a mass audience, who, exactly, should be called a journalist?

In part 2, we examine the broader areas where the Internet has presented the most contentious ethical dilemmas that have wide social implications. These are in the areas of privacy (chapter 4), speech (chapter 5), and intellectual property/copyright (chapter 6). In this section, each chapter begins with a historical perspective and then reviews the legal and social perspectives of those issues as they have evolved over time. We then identify the specific ethical dilemmas and problems confronting online media professionals. Specific case studies are discussed and we explore how these problems were actually resolved by an online media organization that had to face a particular "digital dilemma." The goals of these chapters are to help you learn how to frame these new issues by facilitating your thinking about

1. What is at stake here? What is the dilemma being presented?
2. What do the current laws say? What are current ethical guidelines, if any?
3. What are the slippery new issues not addressed by the law or current ethical guidelines?
4. What are the implications of how some actual online news organizations handled these dilemmas?
5. What should the ethical guidelines be in this circumstance, and why?

In part 3, we focus on three narrower ethical issues confronting on-line media professionals that are of concern specifically to the journalism profession. We begin by providing a quick history of how online journalism emerged and how it has evolved over time. Then we examine the specific ethical challenges to online journalists: maintaining accuracy (chapter 7), performing quality research (chapter 8), and distinguishing editorial matter from advertising (chapter 9). We finish this section with a broader discussion of the Internet and democracy.

As in part 2, each chapter ends by posing *critical thinking questions.* These questions are designed to help you think through the particular dilemmas and to assist you in coming to your own decisions within the context of the ethical considerations and principles at stake. Each chapter also ends with a *summary,* a listing of *key terms,* and *recommended resources* to help you explore more about the topic.

A Few Words About Words

There are some slippery and overlapping terms in this new field, and we think it would be best to clarify up front the words you will find in this book and how we are defining them.

- *Media/mass media:* All of the various types of organizations that collect, filter and disseminate news and information: newspapers, television networks, radio and Internet news sites. (Sometimes public relations is included in this term, but this function is not of primary focus in this book.)
- *Media professional:* Someone whose job involves gathering, collating, analyzing, editing, writing and/or preparing information for publication or broadcast. Media professionals include print, radio, television or Internet-based researchers, reporters, editors, producers or anyone else whose job primarily focuses on the information creation and dissemination process. (As with the example of mass media, this term can also include writers and editors in the public relations field or others whose profession requires communication of information to a wide audience.)
- *New media:* Originally used to describe the Internet as the newest medium for disseminating news and information. Now less often used—see "online journalism."

- *Online journalism:* The preferred term used to describe the practice of journalism over the Internet. An online journalism professional is simply a journalist who works on the staff of an online news site. An online journalist performs online journalism.
- *Online media:* The range of media that conducts their operations primarily over the Internet. An online media professional works in the online media.
- *Online news:* News distributed over the Internet.
- *Online news site:* A media organization that disseminates its news over the Internet.

Note that the terms *online media professional* and *online journalist* will largely be used interchangeably in this book. However, when discussing concerns specifically related to the traditional practice of print journalism, our preference will be to use the words *journalism* and *journalist* for that context.

What about the definition of *journalism* itself? If you look up the word in the dictionary, you'll get a simple definition. For example, the 2000 edition of the American Heritage Dictionary defines journalism as, *"The collecting, writing, editing, and presenting of news or news articles in newspapers and magazines and in radio and television broadcasts."*[1] But if you ask those who work in the profession, you'll get a much more nuanced answer. For example, Jeff Mohl, editor of the Society of Professional Journalists (SPJ) publication *The Quill*, wrote in the May 2002 issue that the definition turns on "the responsibility involved in our work . . . It's intent that defines a journalist."[2]

We actually spend all of chapter 3 discussing this difficult question of who counts as a journalist. This question has been made much more difficult with the rise of self-proclaimed reporters who have created a Web site to tell the world their version of the day's news and broadcast their own views.

PART I

Painting the Larger Picture: Media, Ethical Codes and the Internet Age

CHAPTER 1

Online Media Ethics in Perspective

Chapter Goals:

- Describe why ethics is of special importance to the media profession.
- Review fundamental broader communication ethical principles.
- Analyze the use of formal codes of ethics.
- Review existing media codes of ethics and their application to online journalism.

As a current or aspiring media professional, your chosen field comes with unavoidable, serious responsibilities. Of course, all jobs demand a certain level of obligation to perform quality work and meet clients' and customer's needs. But the media's role in society carries special obligations.

Ron F. Smith, author of *Groping for Ethics in Journalism* (4th ed.), outlines certain goals shared by most American journalists:

1. *To inform the public about incidents, trends and developments in society and government.* Journalists are obliged to gather information as best they can and to tell the truth as they find it. They must be undaunted in their pursuit of truth and unhampered by conflicting interests.
2. *To treat people—both those in the audience and those who are making news—with fairness, respect, and even compassion.* It does journalists little good to strive for the truth if a large number of people do

3

not believe news reports because they do not trust or respect the news media.

3. *To nurture the democratic process.* For people to govern themselves, they must be informed about the issues and the actions of their government. The news media are the chief providers of that information.[1]

On the connection between the media and democracy, it was James Madison who famously said:

A popular government without popular information or the means of acquiring it is but a prologue to Farce or Tragedy or perhaps both. Knowledge will forever govern ignorance, and a people who mean to be their own Governors must arm themselves with the power knowledge gives.[2]

Those who seek to wear the mantle of information provider for their society, therefore, take on a quasi-public service status and must be particularly attuned to the ethical dimension of their choices and decisions.

To many the entire concept of any kind of "media ethics" is an oxymoron. But the fact is that while one can note countless instances of unethical and poor quality practices by specific media outlets, the profession itself has taken the responsibility of doing its job well very seriously. The professional bodies for most of the major media practices have created explicit ethical codes for its membership, and many individual news media organizations have worked diligently to create and implement ethical guidelines for their own employees. And there is no shortage of professional books written for those in the craft that provide advice and instruction on how to do quality, ethical work.

The broadest ethical guidelines outlined by these bodies state the fundamental principles of the craft, including such matters as telling the truth, being accurate, avoiding conflicts and minimizing harm. These are standards that are applicable to media professionals who work in any medium: print, radio, television or the Internet.

But the emergence of the Internet as a new medium for creating and delivering news and information has brought forth a variety of unanticipated and thorny ethical dilemmas. True, you can consult the basic, already articulated ethical principles to try to resolve these quandaries (and we have reprinted the existing codes from several leading media organizations in the Appendix). But it is also true that the guidance on what is the best course of action won't always be clear, since there is little precedent for dealing with these new and unanticipated issues brought about by the digital age.

Sidebar

Ethical Communication
as an Overriding Principle

Ethics, which refers to the principles of right conduct, is a broad scholarly discipline that falls under the study of philosophy and contains many perspectives, subcategories and approaches. In day-to-day life, though, the need to consider the ethical component of a decision becomes practically relevant when we consciously confront a choice or decision and know that our action will have an impact on another person or persons, which could cause them some harm. The ethical dimension becomes important when we have the awareness to want to do "the right thing" rather than just anything at all.

As a media professional, your overriding ethical responsibility is to serve your readers, listeners or viewers. You carry this out by imparting information, a form of communication. So the kind of ethical dilemmas that you will confront can be categorized under the larger umbrella of *communication ethics*.

The determination, though, of what constitutes ethical communication is, as you might imagine, complex. In fact, philosophers over the centuries have analyzed, from a myriad of perspectives, what kinds of communications are ethical and what kinds are unethical. In his lucid and comprehensive textbook, *Ethics in Communication*,* Richard L. Johannesen has identified and categorized several broad "perspectives" that writers, philosophers and observers have taken over the years when examining communication ethics and elucidating their own principles and ethical stances. The perspectives outlined in his book are:

Political Perspectives. This perspective looks at the communications relevant and important in the realm of the *public and society at large.* Johannesen cites several renowned authors here including Karl Wallace, Franklyn Haiman, Thomas Nilsen, Dennis Gouran, J. Michael Sproule, John Orman, F.G Bailey, I.F. Stone, Henry Johnston, Ithiel de Sola Pool and Sidney Hook.

Human Nature Perspectives. The human nature perspective looks at what kinds of communications serve to enhance *human attributes and do not dehumanize.* Specific approaches here include looking at human

---

* Richard L. Johannesen, *Ethics in Communication*, 5th ed. (Prospect Heights, IL: Waveland Press, 2002).

rational capacities and symbol-using capability, the moral imperative of Immanuel Kant, the use of language, epistemological approaches, value judgments, use of rhetoric, and an ethics grounded in the field of evolutionary psychology.

Dialogical Perspectives. Here Johannesen groups those who look specifically at the *attitudes* that people have toward each other and the characteristics important to those who focus on creating a dialogue. Key components of this perspective are authenticity, inclusion, confirmation, presentness, spirit of mutual equality and a supportive climate. Two major schools of thought include Martin Buber's "I-Thou" relationship and Carl Rogers' client-centered, nondirective approach of "unconditional positive regard" for the client.

Situational Perspectives. A situational perspective looks at the particular type of communication *case by case* in order to make a judgment on the ethicality of that specific communication. Among the schools examined here are Edward Rogge's situational perspective; B.J. Diggs' focus on the "contextual character" of the communication; Saul Alinsky's approach that communicators should view truth and values as relative, including the determination of appropriate means and ends; and those who justify obscene and profane words.

Religious Perspectives. A religious perspective looks at communication ethics from the *principles of the religion and the fundamental religious texts.* There is the Christian ethic, which stems from human reflection of God's image; Asian and Middle Eastern religious perspectives such as the Confucian emphasis on facts and logic; the Taoist's values of empathy and insight; Buddhist principles; and a Muslim's emphasis on God's law over human-made ones.

Utilitarian Perspectives. This ethic evaluates the *consequence* of an action or a behavior as the ultimate test and does so by looking at factors like usefulness, pleasure and happiness for the most. Johannesen notes that this perspective is usually combined with one of the other major perspectives.

Legalistic Perspectives. Those who take a legalistic perspective generally take the view that if an action is illegal, it is therefore unethical; and if something is not illegal, than it would be considered ethical.

Feminist Perspectives. The feminist perspective typically looks closely at issues related to balance of power, domination and subordination. This approach looks at relationships in a web and focuses less on the individual as an atomic unit. A well-known writer on feminist ethics, Carol Gilligan, wrote about a feminist ethic of care vs. the traditional masculine ethic of justice.

Professional Ethics

Becoming familiar with these broad approaches to ethical communication is important in order to see how communication and ethics intersect and to understand various approaches to making an ethical decision. They can even offer possible approaches for resolving your own ethical communication dilemmas.

However, although these perspectives can create a foundation for establishing broad principles to live by, for those in a profession who confront practical ethical quandaries on a regular basis, there is also a need for narrower, more practical ethical rules and standards.

These specialized ethical codes and rules are called "professional ethics" and are designed to guide and regulate conduct of members within a particular field. Professional ethical codes exist for all sorts of professions: doctors, lawyers, financial planners, computer programmers, architects, professors, realtors, and, many other practitioners, including of course, the media. Often, the ethical standards for a profession are articulated in a clearly written document, usually referred to as a formal *code of ethics*. Sometimes those formal codes are created by an association in the field for its membership. In other cases, a formal code of ethics is created by an individual organization in order to set out the dos, don'ts and broad principles for its own employees.

Because this is a textbook about ethical dilemmas facing members of the online media profession, all of our discussion about ethics can be said to fall under the category of professional ethics.

Potential Problems with Formal Codes

How are the existing formal codes of value and relevance specifically to *online* media professionals? Before answering this question, it should first be noted that the usefulness of formal codes of ethics is sometimes questioned.

For instance, Jeffrey L. Seglin, a 2001 fellow at the Poynter Institute, a journalism research establishment located in St. Petersburg, Fla., identified various cases of ethical abuses at newspapers that happened despite the publication's written code of ethics.[†] Seglin argues that the problem was

---

[†] "Codes of Ethics: Why Writing One Is Not Enough," *Poynter Online*, December 17, 2002. Retrieved from the World Wide Web: www.poynter.org/content/content_view.asp?id=4697.

that the principles behind the codes were never internalized by the staff, and that internalizing ethical principles requires much more than merely posting or distributing a code. It requires the organization to initiate activities like training, analysis of case studies, group conversations, and active modeling of the stated principles. Seglin noted that a news organization's concern with the bottom line makes it difficult to take the ethical route. Without the extra work that these kinds of internalization efforts require, it is all too easy for a media's staff to simply take the easiest path, which won't necessarily be the ethical one.

Other objections to formal ethical codes are noted by Johannesen, who notes the complaints that such codes are "frequently filled with meaningless language, include 'semantically foggy clichés' and thus are too abstract, vague, and ambiguous to be usefully applied."[‡] Other criticisms focus on whether such codes really work in practice, their lack of flexibility to adapt to new circumstances, and that while formal codes may appear to be universal, they really are not. Some critics are concerned that codes may limit a journalist's First Amendment freedoms. For example, a code could lead to a plan whereby in order to publish, one would have to obtain a license from the government. Others focus on a lack of enforcement, and some detractors take a cynical outlook and worry that such codes are written simply to excuse an organization's current suspect practices. Another objection is that people may incorrectly assume that as long as something isn't prohibited in their organization's formal ethical code then it must be okay.[§]

But despite these real potential flaws, formal codes of ethics have in fact been shown to serve several important purposes, and if implemented correctly, these problem areas can be ameliorated or avoided. Johannesen cites several real potential benefits of creating an ethical code:

- Educating people new to the profession of the field's standards and principles.
- Surfacing potential problem areas ahead of time and providing early guidance on how to resolve ethical problems.
- Facilitating discussion and reflection on relevant and current difficult issues.
- Assisting in creating those character traits desired for those who work in that profession or those working at a particular organization.

---

‡ Johannesen, p. 182.
§ Johannesen, p. 183.

Here Johannesen quotes Karen Lebacqz in her book *Professional Ethics* (Abingdon, 1985) that "a professional is called not simply to *do* something but to *be* something."[#]

- Setting a high ideal worth trying to reach.
- Stimulating discussion among staff on important relevant issues that need reflection.

Other benefits of formal ethical codes, as noted by Deborah Johnson, author of *Computer Ethics,* include creating of a set of expectations that members will live up to the code and sensitizing members to issues they may not be aware of. She also says that codes can help establish public trust and fend off external regulation.[**]

Johannesen writes that even if members of an organization or profession don't live up to a code's standards, that this does not mean that the codes have no value. He states that such codes point to a goal and an ideal that may not always be reached but what members can strive to attain.

To see the formal ethical code for one large media organization, Gannett Newspapers, see the appendix.

A formal code of ethics can be of particular value to professionals who work in the Internet medium. Because this is a new area, still evolving and in flux, in some cases it is not even clear what is a legal or illegal practice. For example, under what circumstances is it okay to download and use another news site's graphic on one's own Web site? A formal ethical code can provide some level of guidance for members in approaching a difficult decision when there is not even a law to consult. Having discussions and debates on questions like this ahead of time and then articulating the organizations' principles can help define what the organization stands for and assist the group in knowing what to do when confronting these kinds of dilemmas.

It is also important to note that there are methods that organizations can use to help ensure that their ethical codes are going to be effective and have an impact. Johannesen says that all language used should be clear and specific—statements need to be spelled out clearly. The code should also address the specific ethical hazards that face members and should help stimulate discussion and reflection among the group so that it can be modified when necessary. It should be developed from input from individuals at all levels of the organization and should not be self-serving. Finally, codes of

---

[#] Johannesen, p. 186.
[**] Deborah Johnson, Computer Ethics (Upper Saddle River, NJ: Prentice Hall, 2001).

ethics should be enforceable and enforced. (For a good example of how one media organization has spelled out in detail its enforcement mechanisms, see the International Association of Business Communicators' code in the appendix).

Existing Professional Ethical Codes in the Media

Online media professionals don't have to operate in a complete vacuum when confronting an ethical dilemma. While this book presents a variety of new and perplexing situations that arise out of working in the digital medium, there are in fact already in place specific professional ethical guidelines and codes that have already been created by professional bodies in the media to assist their membership in dealing with matters of ethics. For example, these media organizations have created a formal code of ethics: the American Society of Newspaper Editors (ASNE), the Associated Press (AP), the National Press Photographers Association (NPPA), the International Association of Business Communicators (IABC), the Radio-Television News Directors Association (RTNDA), the http://www.sabew.org(SABEW), and the Society of Professional Journalists (SPJ).

The codes for each of these entities are reprinted in the appendix (other than the Society of Professional Journalists' code, which is reprinted later in this chapter).

When reviewing those association's codes, it becomes clear that, although each one is different, the basic principles and ethical standards are quite similar. Most of them begin their declaration by stating an overriding guiding principle that provides a larger context for the mission of the organization and then go on to spell out specific ethical dos and don'ts. To illustrate these existing ethical principles and codes we have summarized their general principles, followed by a listing of their specific ethical rules and guidelines.

General Principles

As is made clear in all of the media organizations' ethical codes, the first obligation is to the public. Following are some examples of the overriding principles:

- Enlighten the public.
- Inform the public.
- Provide a fair and comprehensive account of events and issues.
- Be watchful of the public's interest.
- Seek the truth.
- Avoid conflicts of interest.
- Help the public make judgments of the issues of the time.

From these larger principles, a wide range of specific ethical codes are then outlined by these associations for their members. We have summarized them and also grouped them into categories.

- *Personal integrity:*
  Be responsible
  Be accountable
  Fulfill readers' trust
  Keep promises

- *Maintain a "watchdog" function:*
  Reveal public information that is hidden
  Be a constructive critic
  Expose misuse of power/hold those with power accountable

- *Avoid conflicts of interest:*
  Be independent
  Don't accept gifts or favors
  Don't favor advertisers
  Resist pressure
  Respect/honor the public's right to know
  Disclose unavoidable conflicts

- *Seek truth:*
  Be accurate, and correct errors
  Be unbiased
  Be impartial
  Be objective
  Do not mislead or manipulate information

- *Show care for subjects of the story:*
  Show decency, compassion, sensitivity
  Show good taste
  Be particularly careful when covering juveniles
  Treat sources as human beings

Respect the rights of others
Respect subjects' privacy

- *When newsgathering:*
Respect the confidentiality of sources
Question sources' motives
Do not plagiarize
Identify sources

- *When presenting the news:*
Distinguish news from editorial/opinion
Distinguish advocacy vs. reporting
Distinguish advertisements from editorial
Clarify and explain coverage
Be timely

- *Respect community:*
Respect diversity and reflect it in coverage
Do not stereotype
Be sensitive to cultural values and beliefs
Facilitate respect and mutual understanding
Give voice to the voiceless
Provide a forum for the exchange of views
Encourage public to voice grievances
Participate judiciously in community activities, without compromising independence

- *Other:*
Do not write a story for the purpose of winning an award
Report on own organization and practices of other journalists honestly
Preserve freedom of the press and speech

The most prominent of existing ethical codes for media professionals is the one created by the SPJ, because SPJ is the largest of the journalism associations. Here is the full version of SPJ's latest version of its ethical code.

Preamble

Members of the Society of Professional Journalists believe that public enlightenment is the forerunner of justice and the foundation of democracy. The duty of the journalist is to further those ends by seeking truth and providing a fair and comprehensive account of events and issues. Conscientious journalists from all media and specialties strive to serve the public with thoroughness and honesty. Professional integrity is the cornerstone of a journalist's credibility. Members of the Society share a dedication to ethical behavior and adopt this code to declare the Society's principles and standards of practice.

Seek Truth and Report It

Journalists should be honest, fair and courageous in gathering, reporting and interpreting information.

Journalists should:

- Test the accuracy of information from all sources and exercise care to avoid inadvertent error. Deliberate distortion is never permissible.
- Diligently seek out subjects of news stories to give them the opportunity to respond to allegations of wrongdoing.
- Identify sources whenever feasible. The public is entitled to as much information as possible on sources' reliability.
- Always question sources' motives before promising anonymity. Clarify conditions attached to any promise made in exchange for information. Keep promises.
- Make certain that headlines, news teases and promotional material, photos, video, audio, graphics, sound bites and quotations do not misrepresent. They should not oversimplify or highlight incidents out of context.
- Never distort the content of news photos or video. Image enhancement for technical clarity is always permissible. Label montages and photo illustrations.
- Avoid misleading re-enactments or staged news events. If re-enactment is necessary to tell a story, label it.
- Avoid undercover or other surreptitious methods of gathering information except when traditional open methods will not yield information vital to the public. Use of such methods should be explained as part of the story.
- Never plagiarize.
- Tell the story of the diversity and magnitude of the human experience boldly, even when it is unpopular to do so.
- Examine their own cultural values and avoid imposing those values on others.
- Avoid stereotyping by race, gender, age, religion, ethnicity, geography, sexual orientation, disability, physical appearance or social status.
- Support the open exchange of views, even views they find repugnant.
- Give voice to the voiceless; official and unofficial sources of information can be equally valid.

- Distinguish between advocacy and news reporting. Analysis and commentary should be labeled and not misrepresent fact or context.
- Distinguish news from advertising and shun hybrids that blur the lines between the two.
- Recognize a special obligation to ensure that the public's business is conducted in the open and that government records are open to inspection.

Minimize Harm

Ethical journalists treat sources, subjects and colleagues as human beings deserving of respect.

Journalists should:

- Show compassion for those who may be affected adversely by news coverage. Use special sensitivity when dealing with children and inexperienced sources or subjects.
- Be sensitive when seeking or using interviews or photographs of those affected by tragedy or grief.
- Recognize that gathering and reporting information may cause harm or discomfort. Pursuit of the news is not a license for arrogance.
- Recognize that private people have a greater right to control information about themselves than do public officials and others who seek power, influence or attention. Only an overriding public need can justify intrusion into anyone's privacy.
- Show good taste. Avoid pandering to lurid curiosity.
- Be cautious about identifying juvenile suspects or victims of sex crimes.
- Be judicious about naming criminal suspects before the formal filing of charges.
- Balance a criminal suspect's fair trial rights with the public's right to be informed.

Act Independently

Journalists should be free of obligation to any interest other than the public's right to know.

Journalists should:

- Avoid conflicts of interest, real or perceived.
- Remain free of associations and activities that may compromise integrity or damage credibility.

- Refuse gifts, favors, fees, free travel and special treatment, and shun secondary employment, political involvement, public office and service in community organizations if they compromise journalistic integrity.
- Disclose unavoidable conflicts.
- Be vigilant and courageous about holding those with power accountable.
- Deny favored treatment to advertisers and special interests and resist their pressure to influence news coverage.
- Be wary of sources offering information for favors or money; avoid bidding for news.

Be Accountable

Journalists are accountable to their readers, listeners, viewers and each other.

Journalists should:

- Clarify and explain news coverage and invite dialogue with the public over journalistic conduct.
- Encourage the public to voice grievances against the news media.
- Admit mistakes and correct them promptly.
- Expose unethical practices of journalists and the news media.
- Abide by the same high standards to which they hold others.

By reviewing these existing professional codes, we discover the standards that are held in common among these different media associations. All stress the virtues of seeking truth, accuracy, distinguishing news from opinion, preserving the freedom of the press, and avoiding conflicts of interest.

Another more formal examination of media ethics codes was performed in 1999 by Bob Steele, a well-known investigative reporter and the writer of the ethics column at the Poynter Institute, who examined the codes of ethics from 33 different newspapers.[3]

In Steele's survey, the ethical concern covered most often related to conflict of interest. This was followed by reporter/source relationship and then manipulation of photographic images. Those three were followed, in order, by corrections policy, plagiarism, deception and misrepresentation, and privacy. Steele found that the codes gave "considerable attention" to what he called the "timeless values" like fairness and accuracy. Only a few codes addressed the Internet in any fashion.

Applying Existing Ethical Codes
to the Online Medium

If there are already existing agreed-upon standards of ethics for media professionals, do they adequately cover the circumstances that online media professionals are likely to face? Gary Hill, Director of Investigations & Special Segments at KSTP-TV news in Minneapolis, MN, and chair of the Society of Professional Journalist's Ethics Committee thinks so. In his view, there is no need to make any changes to SPJ's existing code to reflect online journalism. He told us in an e-mail interview that "the SPJ code is designed to be medium independent," and that during his three-year involvement with the Ethics Committee, nobody proposed modifications because of any new issues related to online reporting.[4]

It is clear that SPJ's code, as well as the other media codes, was indeed written to reflect the broad values that underlie the most important roles of the media. So, in principle, these existing codes are written to be medium independent. Online media professionals, when confronting an ethical dilemma, should be able to consult an existing ethical code from their professional association, or from their own organization if one exists, to find some guidance.

However, it is also worth observing that codes are written with a certain assumption about the kinds of specific ethical challenges that its members are likely to face, and that many of those challenges have grown out of the nature of the commonly used mediums of print or broadcast.

Furthermore, the specific examples that are provided in the codes to illustrate the kinds of dos and don'ts are designed to offer specific help ahead of time and may therefore relate to a particular medium. For example, the ethical code of the NPPA specifically addresses not altering photographs in the darkroom; and the code of ethics and professional conduct of the RTNDA admonishes against presenting "images or sounds that are reenacted without informing the public" (see the appendix). Precise guidelines that explicitly address challenges from working in a particular medium offer very helpful specific and practical guidance.

There are ethical problems that now face online media professionals that were not anticipated when these earlier codes were written. Given the knowledge of these newer digital quandaries, it would be worth considering if an updated and expanded code could, while not altering the most basic principles, better address these new digital dilemmas and offer examples that are directed specifically to those working in the online media.

What might an online media code look like? One example comes from a successful Internet media site called CNET, which was founded in 1992 and covers the technology and computing fields. According to the firm, CNET employs over 300 editors and journalists and has the largest global audience in its field. CNET created and published an ethical code for its staff, which is reprinted here.

Editorial and Disclosure Statement

One of CNET Networks' core values is integrity. We promise that our coverage will be fair, accurate, and ethical. For us to succeed, users must be able to trust that our news, product ratings, and technical recommendations are unbiased and based on fact and careful analysis.

A media company's challenge is to give its users what they want in a way that leverages the unique power of each medium. Part of that power on the Internet is the ability to bring buyers and sellers together; in doing so, the line between content and commerce must remain clear. Ethical considerations—such as clearly distinguishing editorial content from commerce content—are of the utmost importance so as not to confuse or deceive users.

We have developed guidelines that set out certain minimum standards and fundamental rules in order to uphold our credibility. This policy addresses issues such as the receipt of gifts or services to CNET employees from companies we cover and the trading of shares in companies we cover, to the distinctions between editorial and advertising content.

CNET Networks will disclose business relationships in editorial content whenever relevant

As part of our corporate mission, CNET has business relationships with and investments in and from a number of companies in the technology industry. CNET Networks editors disclose business relationships in editorial whenever relevant, specifically in any content where a CNET partner or investor is a primary subject. Ordinary advertising relationships, content distribution agreements, or service contracts do not require disclosure.

Ads will be labeled as ads

CNET Networks is an advertising-driven business. We generate revenue by selling advertisements in the form of http://www.iab.net/iab_banner_standards/bannersizes.html, sponsorships, special advertising sections, and cobranded services. Advertisements or links that differ from the standards specified by the http://www.iab.net/iab_banner_standards/ bannersizes.html are labeled *advertisement, advertiser links, sponsor,* or *sponsored links.* A number of advertisements do not need labeling because they are obviously ads.

CNET Networks will disclose the business model

CNET Networks receives referral fees from merchants when we match buyers with vendors. CNET does not sell products and does not receive a portion of the vendor's transaction. All merchants in CNET Networks' price listings pay to be listed. Some merchants pay a premium for placement order or prominence.

CNET Networks will not review its own services or services with which CNET has a business affiliation

It is CNET policy not to review its own services or services with which CNET has a business affiliation. These include—but are not limited to— NBCi.com, Email.com, Onebox.com, and Trellix.com. We do review or comment on competitors to these services. In these instances, we publish a disclaimer acknowledging our relationship to a competing product.

CNET Networks employees should avoid any appearance of impropriety

All CNET Networks employees go out of their way to avoid any appearance of impropriety in regard to securities transactions or inside information about any company we cover. CNET Networks doesn't prevent employees from investing in their financial future, but all employees avoid any activity that might be perceived as speculation or misuse of information obtained at work.

Employees may not accept gifts of value

Editorial employees are required to refuse offers that could be misconstrued by the giver or give the impression that CNET Networks is beholden to a company. This includes the acceptance of gifts, travel, and tickets to entertainment or sporting events.

Labeling content from external sources

CNET Networks labels content to indicate the source in cases where the information published was not generated by CNET's editorial staff.

Corrections

We are committed to correcting quickly any errors in fact or interpretation. No error brought to the attention of an editor or reporter will be allowed to remain in a CNET Networks article. Everything from a mis-spelled name to a gross material error will be corrected in a story's orig-inal file so that no inaccuracy remains in our permanent archive. We will also provide clarifications of stories that contained no clear error in fact but whose wording or headline is open to misinterpretation.

Employees may not keep products or services

Editorial employees often solicit products for review. To avoid the ap-pearance that the receipt of these products constitutes gifts, no employee can keep any hardware for more than six months. Free accounts to online services will be closed after the review is published. Software does not need to be returned but cannot be resold or used for commercial pur-poses.

Linking to external sites

CNET Networks may link to non-CNET content that could be considered offensive by certain readers. The use of a hyperlink never constitutes an endorsement of a company, product, or opinion. The use of the hyperlink does not imply that CNET Networks agrees with the content being linked to and provides no guarantee that the content contains accurate informa-tion. CNET Networks editors are under no obligation to link to any site.
Last updated October 10, 2000.[5]

It is interesting to note that several of the key principles articulated in CNET's code are the same as those created by media organizations for their non-online membership, e.g., be fair, accurate, and provide unbi-ased coverage; distinguish editorial from business; don't accept gifts; perform necessary disclosures, and make corrections. But note too that portions of CNET's code discuss matters of concern only to those who work in the online media: labeling advertisements or links that are not clearly ads; and creating policies regarding linking to external sites. Such examples related to the medium can be helpful to the online me-dia professional.

Weblog Ethical Codes?

CNET, though it operates online, is still, at heart, a media organiza-tion. Another form of online media that diverges from traditional jour-nalism is Weblogs, those informal one-person forms of daily diary and linking journalism (discussed in chapter 3). Should bloggers be expected to follow the traditional journalistic codes of ethics? Would they be covered by an ethical code designed for online journalists? Or do they need their own special code of ethics altogether? Chapter 3 goes into some depth exploring the nature of the relationship between Weblogs and traditional journalism, but here it is worth briefly noting that a distinct code of ethics has indeed been proposed for the Weblog community.

Rebecca Blood is author of a book on Weblogs[6] and a Weblogger herself. In her book's section on ethics and Weblogs, Blood explores how the mission of the Weblogger varies from that of the traditional journalist, noting that bloggers are the "mavericks of the online world . . . beholden to no one, Weblogs point to, comment on, and spread information according to their own, quirky criteria." She writes that "two of their greatest strengths are their ability to filter and disseminate information to a widely dispersed audience, and their position outside the mainstream of mass media." She goes on to propose six ethical rules that she believes ethical Webloggers will abide by:

1. Publish as fact only that which you believe to be true.
2. If material exists online, link to it when you reference it.
3. Publicly correct any misinformation.
4. Write each entry as if it could not be changed; add to, but do not rewrite or delete, any entry.
5. Disclose any conflict of interest.
6. Note questionable and biased sources.[7]

Respecting the Values of a Global Audience

When an ethical code is created, it normally reflects the values of the culture and country of the persons who developed that code. But values and ethics, of course, differ quite a bit among cultures, not just within one country, but dramatically between countries and regions of the world. For example, in the United States, plagiarism is considered an unethical act, akin to theft; but in certain other parts of the world, it is not considered a transgression of any type at all.

A fundamental characteristic of the Internet is that communication and publishing are available to a worldwide audience. This global reach of the Internet has already created several thorny and hard-to-resolve legal issues. For example, can a person or organization in one country be sued for libel by a party in another country, based on the home country laws of the plaintiff? Does a site in one country need to comply with restrictions on what is allowed to be sold in another country? The international communications capability of the Internet also has ethical ramifications. Should this global nature of the online media play into the development of an ethical code? Whose culture and country's values should the code reflect—*both* the senders and the receivers? But receivers of information from the online media are likely to be from *many* countries around the world. What can be done to address this? Could one develop something like a transnational code of ethics that reflects the values of all people, in all cultures worldwide?

This question raises the broader issue of whether or not there is such a thing as universal values, and therefore a universal ethic of communications. Richard L. Johannesen quotes one author who says that such an ethic implies that any kind of universal value would be applicable "for all persons, times, or viewpoints" and that they reflect "common" human traits.[8]

Johannesen describes the works of various authors who have proposed an approach to developing such an ethic. One of the most familiar approaches is the Golden Rule—do unto others as you would have them do unto you. And, in fact, Johannesen writes that a variation of this rule is found not just in Christianity, but in some form or another in all of the world's major religions, and in some smaller ones as well. A more refined version of the Golden Rule was developed by Milton Bennet, who rephrased it as "do unto others as they themselves would have done to them" to avoid the problem of assuming that the other wants to be treated just as you would like to be treated, which may not always be the case, particularly among cultures and countries.[9]

While it may be daunting, and perhaps ultimately impossible, to truly come up with a workable and practical general code of communication ethics to apply to everyone worldwide, it is easier to try to discern a universal professional code. In this case, it means trying to find out if there are common values that media professionals hold around the world. To try to discover this, it is instructive to review existing ethical codes created by journalism bodies from various countries. A selection of these from countries in Western and Eastern Europe, as well as others from Latin America, Asia and Africa are included in the appendix. (More international journalism ethical codes than are reprinted in the Appendix can be found on the home page of the International Journalists Network. See www.ijnet.org/code.html).[10]

By browsing these codes, it becomes clear that, interestingly, virtually all of the primary values of the U.S. ethical codes are replicated in these other countries' codes. Specifically, they all stress the importance of seeking truth, being accurate and fair, avoiding conflict of interest, and showing compassion to subjects, and even include some of the narrower provisions such as distinguishing advertising from editorial and opinion from news and making corrections. This may be reassuring for those who wonder if media in other countries follow a similar set of values as do those in the United States.

It's also interesting, though, to note those codes that seem unfamiliar or even a bit unusual. Here are a few examples[11]:

- "In cases of accidents and catastrophes, saving the injured and victims has the priority before the right to inform the public." (Slovenia)
- Engaging in intelligence work damages credibility and trust. (Germany) Similarly: A journalist who serves as an informer to an intelligence service discredits the profession of a journalist and the function of journalism. (Slovenia)
- It is irreconcilable with the code to praise criminal, terrorism, violence and inhumanity. (Slovenia)
- Physical and psychical diseases or impairments are an inalienable part of personal privacy and as a rule cannot be the subject of journalistic reporting when names are mentioned. (Slovenia)
- Any kind of warmonger activity or engagement in igniting conflicts among different nations is irreconcilable with the code. (Slovenia)
- Journalists should not have to write or do anything contrary to his/her own convictions. (Norway, Finland)
- Avoid sensationalism in reports on medical issues "which could arouse baseless fears or hopes in the reader." (Germany) Here Germany's code gives a specific example: "Early research findings should not be presented as though they were conclusive or almost conclusive."
- In the case of accidents and disasters, the press shall bear in mind that rescue operations for victims and persons in jeopardy take precedence over the public's right to be informed. (Germany)
- Faithfully reproduce contents of documents. (Germany)

- Physical and mental illnesses and disorders fall within the confidential sphere of the person concerned. Out of respect of the privacy for that person and his family, the press should refrain from publishing names and photographs in such cases and avoid using disparaging names for medical conditions or medical institutions even if such names are found in common parlance. (Germany)
- Avoid publication of criminal memoirs from perspective of perpetrator. (Germany)
- Incompatible with press responsibility to publish text or pictures whose form or content could deeply offend the moral or religious sensibilities of a particular group of persons. (Germany)
- Provide detailed methodological data on survey polls, including sponsor of poll. (Germany)

It is interesting to note that several of these codes specifically address the ethics of court reporting, laying down stricter standards than those found in most U.S. media organizations. Norway's and Denmark's codes warn against publishing information that may obstruct a case or pronouncing that a suspect or an accused is guilty. Germany's Article 13 states that: "Reports on cases under criminal investigation or subjudice must be devoid of all preconceived opinion. Before and during such proceedings, therefore, the press shall avoid making any comment in the heading or body of a report which could be construed as partisan or prejudicial to the issue. An accused person must not be presented as a guilty party before legal judgment has been pronounced."[12] And Finland instructs journalists to "only mention names in relation to a crime if considerable public interest is served."[13]

Another area where there appears to be more sensitivity than in the United States are in matters that relate to crime and accidents. Denmark and Finland discuss the need to avoid photographs in bad taste, such as gruesome accidents. Norway and Denmark also single out suicide or attempted suicide as causes of accidents that should not even be mentioned, and Germany's code advises the use of restraint in these cases.

What this review suggests is that while the basic principles of the U.S. media are generally agreed upon in many countries around the world, there are also unique special sensitivities and cultural prerogatives that may cause some surprises for online media professionals who serve a global audience. It is difficult to create an ethical code that reflects and respects the differing values of one's readers who share the *same* national viewpoints. It would be even more challenging to create one that respects the values of a worldwide readership with a very divergent set of values.

Summary

A media professional takes on unavoidable and serious professional ethical responsibilities in carrying out his or her work. The major media associations, such as the Society of Professional Journalists, as well as individual news organizations, have created formal "codes of ethics" to guide and regulate their members or employees in what is considered ethical and unethical behavior. Although the value of formal codes of ethics is sometimes questioned, if they are instituted correctly, they can serve an important function.

The major media organizations' codes of ethics have many principles in common. Among them are to inform and serve the public's interest, to seek the truth, to avoid conflicts of interest, to be fair, to be accurate, to serve as a watchdog and a check on abuses of power, to treat subjects of stories with decency and compassion, to keep editorial separate from advertising, to serve the interests of one's local community, and to help protect free speech.

These existing media codes can be adapted and updated to address some of the novel and unanticipated ethical dilemmas facing media professionals that work online. One online news organization that has created its own code of ethics is CNET. A separate, more informal, one-person type of online journalism is called Weblogs, and a code of ethics has been proposed specifically for those online communicators.

Since readers of an online news site may originate from anywhere around the world, they will have widely differing values; therefore, another special concern for online media professionals is whether it is possible to identify any universal values and then integrate these into an ethical code.

Critical Thinking Questions

- Identify three principles outlined in the SPJ code of ethics that you think are honored the least in actual practice by traditional print and broadcast journalists. Are media professionals that work in the online medium likely to do a better job in complying? Why or why not?

- Based upon the criticisms of formal codes of ethics, what are some alternative methods that an organization could use to create a high ethical standard for its employees?
- Explain whether or not it is necessary that a new code of ethics be drawn up to specifically address the practice of online journalism.
- Should an online media site try to consider the values of readers in other countries when creating its code of ethics? Do you believe there are universal values that apply across cultures and nations? If so, what are they? How would a news organization respect those values?

Key Terms

Code of ethics
Communication ethics
Ethics
Professional ethics
Universal values

Recommended Resources

Books

Johannesen, Richard L. *Ethics in Human Communication*, 5th ed. (Prospect Heights, IL: Waveland, 2002).
Johnson, Deborah. *Computer Ethics* (Upper Saddle River, NJ: Prentice Hall, 2001).
Smith, Ron F. *Groping for Ethics in Journalism*, 4th ed. (Ames: Iowa State Press, 1999).

Associations and Institutes

The Poynter Institute. Web site: www.poynter.org.
The Society of Professional Journalists (SPJ). Web site: www.spj.org.

CHAPTER 2

The Role of Media in Democratic Society

Chapter Goals:

- Provide a historical background on the evolution of the U.S. press.
- Identify emerging journalism movements and their impact on the profession.
- Describe new movements that attempt to improve journalism's role in a democracy.

This chapter will consider the relationship between the media and democracy. The primary issues to be addressed are the historical role of journalism in the formation and growth of American democracy, the ways in which technology and economics have changed the processes and products of journalism, and the current and future challenges for journalism in the digital media era.

Before going further it's important to point out that professional journalism, as we know it today, did not exist during the 17th and 18th centuries when the seeds of democracy were first planted, or during the decades leading up to the Civil War. It wasn't until the late 19th century that the daily gathering and reporting of news came to be considered a professional endeavor, requiring a combination of skill (interviewing, researching and writing), education (high school and college-level), and social responsibility (ethics).[1] This is not to say that *journalism* (of a sort) did not exist prior to the 1890s; rather it serves to highlight the fact that for a big chunk of American history "doing journalism" often wasn't a reporter's sole occupation, nor was it considered to be a respectable pursuit.

This raises some intriguing questions. In the absence of professional journalists, who published and wrote for the colonial and early American press? What sorts of backgrounds did they have? How accurate, fair or balanced was their writing? How did they attract readers and gain financial support? And, perhaps most interestingly, if democracy could emerge and grow in the absence of professional journalism, why is there so much emphasis today on the importance of professional, ethical journalism? What changes in politics, culture and economics led to professional journalism in the first place—as well as other forms of professional communication such as public relations—and how did it become so tightly linked to democratic governance?

It will be impossible to surface conclusive answers to all these questions, but we can attempt to shed a little light on them within the framework of this book. By looking back at the historical relationship of the American press and democracy, and the transformation of the press into the mass media, we might be able to identify more clearly some of the forces that guided the transformation and highlight some ethical perils that confront professional communicators in the age of multi-media, digital dissemination and interaction.

Journalism and the American Revolution

The Press and Political Dissent

While the press in colonial America was not staffed by what we would call professional reporters, it was in many cases operated by rather able public thinkers and writers with broad intellectual interests. Benjamin Franklin is a perfect example. He was one of the most noted publishers of his day, yet publishing was only one of many interests he pursued. Franklin and his brother James, who published a prominent newspaper in Boston, could be called public intellectuals and political activists. Benjamin Franklin used his newspaper (and books) to disseminate information, promote scientific discoveries, and challenge the policies of the English monarchy, liberally mixing news with opinion and commentary.

Other prominent publisher/activists followed this model, openly challenging the crown and colonial governors, often using a pseudonym or writing anonymously. Political rhetoric that was too radical for the papers would often be published in pamphlet format. Thomas Paine's *Common Sense*, a fiery political essay that argued for political independence, is one of the most notable examples of advocacy journalism in American history. It stoked the flames of revolution and was hugely popular, selling an estimated 100,000 copies.[2]

Also in years before the Revolutionary War, the press actively promoted resistance to British trade policies, notably the Stamp Act and the Townsend Act, which they argued were acts of "taxation without representation." In 1765, the English governor of New York was reported to have complained that the colonial press had used malicious lies to excite the people to commit acts of disobedience, which was largely true. Publishers had used inaccuracies and exaggerated reports to whip up opposition to British laws, though they occasionally muted their rhetoric under the threat of being sued for seditious libel.[3]

The Press and National Identity

Though the political focus of much of the colonial and revolutionary press was anti-British and pro-independence, there were numerous papers that supported British rule, illustrating the sharp political divide in the colonies. In addition, the layout and typography of the press was often borrowed from popular British, and occasionally French, journals that circulated in America. Mitchell Stephens, a journalism historian at New York University, reports that

> To be an American newspaper printer in 1760 in a port city, consequently, required poring through as many English, French and other foreign newspapers as it was possible get off the ships. Those papers were generally more advanced in reporting methods, in typography, in design and in writing style than American newspapers. They remained so into the nineteenth century. Our editors, consequently, borrowed more than stories from them. European newspapers provided the forms from which American journalism was cast—and recast.[4]

In addition to reprinting news from Europe and domestic political essays, the press also disseminated news about piracy on the high seas, local disasters (such as fires), counterfeiting, robberies, obituaries and religion. Advertising was limited and consisted primarily of government announcements and lists of goods arriving at the nearest port.

By borrowing and recasting the style of European journalism, the more radical American publishers were supplying fellow colonists with news from their home countries and at the same time helping to plant the ideas of revolution and independence. (In reciprocal fashion, many in Europe, especially France, borrowed political ideas from the Americans.) Thus the press helped foster something of an uneasy parent-child relationship between colonists and England that was animated by the acceptance of some inherited traditions and values (English common law and the dissemination of printed news) and the rejection of others (monarchy and feudal social relations). In this way, a new national identity was being formed. Gradually, as the papers began to refer to fellow colonists as "Americans," readers began to feel as though they were part of a new political and cultural community.

As important as the press was, dissemination of printed news and political opinion must not be given too much credit for the exchange of news and the rise of revolutionary sentiment. Many colonists could not afford to purchase newspaper subscriptions, while others were illiterate. As such, face-to-face dialogue was just as critical to the spread of news and opinion. Furthermore, among those who could read and afford papers, oral conversation and debate helped give meaning to words on the page—to put news and ideas into context, to relate them to personal experience. Media theorist Harold Innis has also suggested that oral communication is potentially the most democratic medium in that it is not easily monopolized. The development of print media, Innis argued, tended to place the emphasis on advances in technology, which could be monopolized by those with political and economic power, who could then control the flow of information and manipulate public opinion. Thus Innis saw the oral tradition as a necessary counterweight to the "monopolies of knowledge" created by mechanical communication. Democratic culture, he believed, was best nurtured when there was a balance between the spoken and printed word.[5] This argument has been advanced by several contemporary media scholars and critics, most notably James W. Carey of Columbia University, who argue that modern journalism should disseminate verifiable information and serve as a channel through which citizens from diverse backgrounds can engage in the "conversation of our culture."[6]

The Press as Public Sphere

The fact that people operating outside the orbit of government—even in open resistance to it—were exchanging news and opinion both orally and in print is evidence that a "public sphere" had emerged in America. This elusive concept has been best articulated by German philosopher Jurgen Habermas, who identified the public sphere as a web of loosely structured social relations, news journals and physical places—salons, coffee houses, taverns, town squares, meeting halls, and so forth— through which private citizens, primarily from the growing merchant class, could meet as equals to debate the major issues of the day without being censored by political or economic powers.[7] The public sphere, in short, is where public policy is debated and public opinion is formed, or as one scholar has put it, the place "where democratic culture happens."[8]

In this light, we can see the connection between the press and democracy. The existence of an independent press along with physical spaces for public debate, according to Habermas, were critical to the success of democratic revolutions in America and Europe (e.g., the French Revolution).[9] Through the process of political debate in print and on the streets citizens could criticize the institutions and policies of the feudal state and articulate a vision of the constitutional governments that would follow its collapse. As the new governments emerged, the public sphere served as a check, or a countervailing power, against new forms of political and economic power. This is not to say that the activities in the public sphere by themselves determined the course of revolutions and the content of subsequent social policies. Instead, it is to point out, as one historian has, that the public sphere "contributed to the spirit of dissent found in a healthy representative democracy."[10]

As an organ of dissent, the colonial press played a part in making the American Revolution possible. The public sphere, then, is where we find the intersection between journalism and democracy. As Stephens indicates, the early conception of a journalist was that of a public "minister," or servant, who would bring issues before the public, interpret its wishes, and circulate its edicts.[11] In keeping with Habermas' model, the public sphere, and hence journalism, lose its democratic character when either the government or large commercial interests, or a combination of both, control the flow of information and dominate the formation of public opinion.

Journalism in Early America

In summarizing the history of the American press, Harold Innis wrote, "The active role of the press in the Revolution was crowned by a guarantee of freedom under the Bill of Rights."[12] This, indeed, was the case as the framers of the Constitution, led by James Madison, worried that unless the press was protected from government interference, a dominant political faction or party would use its legislative power to censor opposing views. In this way, the Bill of Rights implies that democratic self-governance is not possible without a free, independent press.

The Partisan Press

By 1800, there were an estimated 200 newspapers in America, including 24 dailies. Many of these were launched by publishers aligned with either of the two major political parties at the time, the Federalists and the Republicans, often with the direct financial support of party leaders.[13] Thus the press continued focusing on politics, though it shifted from calls for revolution to debates about the structure and policies of the new government from a decidedly partisan point of view. And partisan debate often deteriorated into vicious name-calling, spreading of false rumors, and other forms of personal attack. Additionally, numerous press operators would print favorable or unfavorable views about a particular party or politician depending on who offered the highest bribe.[14]

Aside from overt bribes, subsidies typically came from government printing contracts—awarded to papers supporting those in charge of the federal purse strings—contributions from party leaders and subscriptions. Advertising, in the form of announcements regarding the arrival of commercial goods in port, was not a major source of revenue, though it typically appeared on the front page of many papers right next to editorials promoting the publishers' political views. Moreover, the press was not seen as a major commercial industry; owners didn't expect huge profits or much profit at all. Many wealthy party supporters were, in fact, willing to accept considerable financial losses so long as their ideas were disseminated. The early press also benefited from government-supported expansion of the postal infrastructure and low postal rates supported by many in government.[15]

The number of newspapers continued to expand rapidly in the early 1800s, from 359 in 1810 to 852 in 1828.[16] By the 1830s, visiting French nobleman Alexis de Tocqueville would remark, "[n]othing is easier than to set up a newspaper, as a small number of subscribers suffices to defray the expenses." Tocqueville also recognized that newspapers were not just a product of big city culture: "In America, there is scarcely a hamlet which has not its [own] newspaper."[17] In fact, many cities and towns had several locally operated papers, representing the views of the major parties and regional factions within them. In addition, political radicals (abolitionists, suffragists, labor organizers and religious leaders) operating outside the party system launched journals and published pamphlets that challenged government policies often on moral, rather than utilitarian, grounds. Consistent with the decentralized, politically oriented ownership, Tocqueville observed that the press either attacked or defended the government "in a thousand different ways."[18] This would seem to indicate that the press was playing something of a watchdog role—a partisan, often irresponsible watchdog, to be sure, but a watchdog nonetheless. This function was vital to democracy, Toqueville believed, especially during a time of rapid industrial expansion, which could give rise to a "manufacturing aristocracy" that might dominate political and economic affairs. So long as there was a large number of news outlets with diverse ownership and low barriers to entry (low production and distribution costs), Toqueville suggested that the press might serve as a countervailing force against political corruption and concentrated economic power.

In a sense, the dominant, partisan press was "independent" from the federal government in that it wasn't overtly owned or controlled by Congress or the president, but it was clearly attached to the two wings of the political establishment through party contributions and federal contracts; and it benefited greatly from government efforts to build communications infrastructure. Despite these contradictions, according to Stephens, the press as a whole helped serve the growing public by disseminating "crucial information about the government," stimulating political discourse, and showing that "leaders were fallible" while "truth was subject to dispute."[19]

The Press as Commercial Mass Medium

Major shifts in American politics, economics and culture during the 19th century coupled with rapid technological advances fueled the transformation of the press from a decentralized institution supported by subscribers and political partisans to a major information and entertainment industry supported by advertising. Spatially, the territory of the United States expanded, causing the population to spread out, which, in turn, necessitated the construction of coast-to-coast communications infrastructure. Democracy was also expanding—though much more slowly and erratically—due to the enfranchisement of whites who did not own property and the abolition of slavery, though women were still denied the right to vote.

The Penny Press

During this time of expansion, the monopoly of the partisan press in large cities was being broken by a new generation of publishers who sought readers from beyond the political elite. Drawing on the British practice of selling cheap, general interest newspapers to industrial workers and other city dwellers, Benjamin Day launched the *New York Sun* in 1833 announcing that the purpose of his paper was "to lay before the public, at a price within means of everyone, all the news of the day, and at the same time offer an advantageous medium for advertisers."[20] The paper was initially priced at two cents and later lowered to a penny, as opposed to the six cents that partisan and "mercantile" (business) papers typically cost. Subscriptions could also be purchased for about three dollars a year, a price that was now within range of most low-wage workers' budgets. Papers could be sold this cheaply because advances in printing technology and the availability of cheaper paper had dramatically lowered production costs. In addition, a large readership could attract advertisers from the merchant class, who could promote their wares for 30 dollars a year. Day also borrowed another innovation from the British press that helped spread his paper to the masses; he distributed the *Sun* to carriers who fanned out across the city, delivering the paper to subscribers and selling copies to passersby.[21] Thus the first paper routes as well as the street vending of news were born.

With regards to content, Day filled his pages with a mix of human-interest stories, local crime reports and news about ships arriving in New York or leaving for Europe. When public affairs were covered, the *Sun* was far less partisan than its more expensive rivals, though Day wasn't above promoting issues and politicians to his liking. His formula was a huge success. Within two years, the *Sun* had the highest circulation figures in the United States.

Following close behind Day was Robert Gordon Bennett, another innovative publisher who started the *New York Herald* in 1835. Though Bennett was an ardent supporter of then president Andrew Jackson, he announced that his paper would shun "all party, all politics" as an editorial guide in favor of "good, sound, practical, common sense."[22] Bennett didn't quite keep his word, regularly using his paper to support Jackson's policies. But, unlike the overtly partisan papers that relied directly on government contracts or party patronage, which forced them to stick with their benefactors or risk going out of business, Bennett had the freedom to put his political support wherever he wanted. His highly profitable paper allowed him to back politicians and issues without fear of losing contracts or political favors.[23] It is also worth noting that even though political reporting was not as dominant in the penny press, the fact that these new papers reached so many people may have, as Stephens suggests, played a "significant role in informing the poorer classes," which helped "involve them in the political process."[24]

Aside from his low prices, populist politics, and gossipy, high society pages, Bennett significantly altered journalism in other ways. In addition to hiring a European correspondent, establishing a Washington Bureau, and posting reporters in other major U.S. cities, he helped introduce the practice of investigative reporting. In 1836, Bennett produced several lengthy stories describing his visits to the scene of a brutal murder. He also conducted interviews with a prominent witness; on one occasion he even printed a transcript of their conversation. The stories stirred New Yorkers and circulation of the *Herald* increased. Through his reports on the case, Bennett had gone beyond official sources and records—which his competitor the *Sun* had relied upon—to search for information independently.[25] And while he wasn't on the trail of a corrupt politician or uncovering a major government scandal, Bennett had created a method of reporting that would later be widely used by journalists in later decades to expose corporate malfeasance and bust political machines.

Despite the huge success of Bennett's editorial model and reporting style, some publishers remained unimpressed. In 1841, Horace Greeley, an abolitionist whose politics were much more radical than Day's or Bennett's, launched the *New York Tribune*, which also sold for a penny. He disdained the sensational crime reporting and gossip of the *Sun* and the *Herald* and positioned his paper as a fierce public watchdog and crusader for social justice. He editorialized against slavery, capital punishment, and the slum conditions of New York's poor, mostly immigrant neighborhoods. Greeley also actively supported the prohibition of alcohol, labor unions and westward expansion ("Go west, young man, go west!"). He was also noteworthy for hiring gifted writers and intellectuals as columnists, including a German radical named Karl Marx. As with Day and Bennett, the success of Greeley's paper made him a very rich man, though he was known to have distributed company stock to some of his employees and make generous donations to the poor.[26]

The explosive growth of the penny press—first in New York City and then in other cities—and the varied editorial approaches employed by its publishers, plus the continued existence of partisan papers and radical suffragist and abolitionist journals, indicates that the news consumer of the mid-19th century had a fairly diverse array of affordable news outlets from which to choose. In a broader sense, the increasingly competitive publishing business combined with a diverse, decentralized ownership structure may have prevented any one outlet, or editorial voice, from dominating political and cultural discourse—there was a democratic "check and balance" that prevented concentrated media power. Thus journalism of this era, while certainly leaving much to be desired, appeared to serve as an important source of information and public debate.

The Telegraph and Industrialization

The prominent changes made to journalism by the penny press might seem insignificant when measured against the changes brought on by rapid industrialization and the emergence of new technology between the 1840s to the early 1900s. Like almost every other enterprise in America, the press would be reorganized under the industrial model, which emphasized corporate hierarchies, the strict division of labor, mass production, and centralized distribution over the more communitarian approach to manufacturing that dominated the colonial and mercantile periods in American history.

For the press, the most significant new technology was the telegraph, which was developed by Samuel Morse in 1844. This new medium was important for two reasons: First, telegraph lines were built along the same public rights of way being used by the railroads. Once again, the government was subsidizing the construction of communications infrastructure, but unlike the case of the postal service, the telegraph was privatized, giving rise to the country's first media monopoly—Western Union.[27] Second, the telegraph conquered space like no technology before it.[28] Information could now travel faster than transportation. Early on, much of the information sent by telegraph was delivered to local commodity markets, which were being reorganized into national exchanges. These markets required vast amounts of information, such as warehouse receipts on agricultural goods, which had to move faster than products.[29]

Taking note of the telegraph's capability were newspaper publishers, including Bennett, Greeley and other pioneers of the penny press. In 1848, several major newspapers joined forces to create the Associated Press, which became the country's first wire service, supplying local and regional presses with the same news from far-off places. During the Mexican-American War, the *New York Herald* and the *Tribune* ran virtually identical stories transmitted by the telegraph.[30] Publishers soon began to think more in terms of the quantity of the information they could gather from various places and transmit to other places rather than the meaning or the function of the information. News was becoming another industrial commodity that needed to be produced quickly and distributed widely. This is a subject that media critic Neil Postman addresses in his book *Amusing Ourselves to Death*. "The principle strength of the telegraph," he writes, "was its capacity to move information, not collect it, explain it or analyze it."[31] Thus the telegraph was a technology of dissemination *par excellence*, and as such, it conformed to Harold Innis' view that the "effect of modern advances in communications was to *enlarge the range of reception while narrowing the points of distribution.*"[32] This contributed to the centralization of newspaper ownership as the bigger publishers began forming partnerships with each other and began buying up smaller papers. The ultimate irony, according to Postman, might be that while the telegraph and industrialization gave more people access to more news, the news had less connection to their personal experiences. Postman argues that the context behind the news gradually began to disappear as stories became organized around facts rather than ideas or a chronological summary of events.

By the 1880s, the decentralized partisan and penny presses were being displaced by a more centralized, commercial newspaper system that followed the logic of the telegraph and the commercial marketplace. Accordingly, the content of the press underwent dramatic changes. Initially, many large papers adopted the editorial model popularized by Day and Bennett, steering even further away from politics. Political editorials and opinionated essays were dropped from the front page in favor of the latest news.[33] A study conducted by journalism historian Ted Smythe indicates that the percentage of space devoted to politics in a group of highly regarded urban newspapers "declined from 50 percent" during the pre-Civil War era to "19 percent in 1897, while space devoted to crime and courts, accidents, society, and leisure activities . . . increased three-fold or more."[34] This style of journalism was less likely to upset readers with strong party affiliations. It helped attract wider audiences, which could be used to lure advertisers eager to promote new mass-produced products. By 1920, advertising would account for two-thirds of all newspaper and magazine revenue.[35]

As it enriched the new commercial press owners, the rise of advertising also had the effect of weakening much of what remained of the radical political and labor press, which typically depended on subscriptions for revenue. With a bounty of ads, a publisher could afford to sell papers for less than the cost of production. Popular noncommercial papers found it hard to stay in business and either had to raise prices to keep up with rising production costs; change their content in order to recruit more consumers, and hence advertisers; or close up shop.[36] In this manner, some media scholars and critics have argued that the market served as a filter, or censor, gradually weeding out dissident publishers and radical voices. Commercialization of the press also changed the nature of competition. It became less a direct battle waged through the price system as it had been during the days of the penny press and more a battle for and through advertising. Complicating matters was an increasing number of advertisers who wanted to attract privileged audiences that had more purchasing power, while others wanted to sell their wares to the widest audience possible. Therefore, commercial publishers had to decide whether they would skew their reporting toward the mass audience, which could generate large circulation figures, or go after the smaller but wealthier business class.

The "New Journalism"

Media historian Michael Schudson reports that in the 1890s, two styles of commercial journalism would emerge: (1) the "story," or entertainment, model employed by newspaper moguls Joseph Pulitzer and William Randolph Hearst; and (2) the "information" model used by *The New York Times*[37] Often called sensational or "yellow" journalism (named for the "Yellow Kid" cartoon), the Pulitzer/Hearst approach was characterized by self-promotion, bold headlines, illustrations, lots of advertisements, and news about crime, sensational court trials and high society.

It might not be fair to lump Pulitzer and Hearst too tightly together, according to Stephens, who argues that Pulitzer cared far more for accuracy than Hearst and was more consistent in his progressive politics than Hearst, who took more conservative positions as the Progressive Era waned.[38] Both publishers sought large audiences and their reporters were quick to chase the most sensational stories, which could be told in a narrative style that would engage the reader by connecting him or her to the experiences of the story's subjects. Hearst's and Pulitzer's papers typically drew large, working-class and immigrant readers. It's also worth noting that Pulitzer is credited with helping introduce the "tabloid" style newspaper. Unlike traditional newspapers, tabloids were folded like a magazine and may have been designed to meet the needs of commuters who were beginning to ride cable cars and other forms of public transportation by 1900.[39]

Contrary to the Hearst/Pulitzer model, the "information" approach practiced by *The New York Times* was aimed at upscale readers. These were professionals, such as doctors, lawyers, executives and managers of newly powerful corporations; they expected the paper to deliver verifiable facts in a nonfrivolous manner—"all the news that's fit to print". Gradually, the "objective" or professional school of journalism, aimed at the middle and upper-middle classes, that began to develop in the 1920s and '30s would replace this model.[40]

Schudson also indicates that elite readers of the *Times* may have been attracted to more than the content and style of presentation; the paper was also seen as a "badge of respectability."[41] To read the *Times* meant one had achieved a certain status in society and was expected to be involved in important political, economic and cultural affairs. Compared to the politically active readers of the *Times*, Schudson notes, the readers of Pulitzer's entertaining *New York World* were "relatively dependent and nonparticipant" in political and business affairs.[42]

Some media critics have suggested that the implication behind this editorial split is a sort of class prejudice that still exists in the media today. It holds that people in the wealthier political class need accurate business news and a realistic picture of world events and trends if they are to successfully manage major political and economic institutions. Meanwhile, the rest of the population is to be treated as passive spectators, except when it comes to purchasing the goods advertised in the papers. Participation at the point of purchase was then and is now enthusiastically encouraged.[43] It is worth asking whether the sensational journalism of the 1890s was so popular because readers demanded it or because it was so heavily promoted and distributed in their neighborhoods.

Several scholars who study advertising point out that, starting with the commercial press in the 1890s, messages promoting the consumer lifestyle began to appear. This was a marked change from earlier periods in American history in which citizens were often discouraged from blatant consumerism. The cultural emphasis had been on hard work, discipline and frugality; the role of the worker was to produce, not acquire. However, as industrial capitalism reached a crisis of overproduction—partly due to the strict division of labor and long work hours—inventories soared. Unless new markets were created, profits might level off or drop. Advertising, then, became much more important as a tool for selling unused products. And the popular press became the perfect vehicle for delivering ads to consumers.

The way in which ads convinced workers to raise their consumption level is rather sophisticated. To put it simply, industrialization had the effect of disrupting old social relations and patterns of communication. People who had once produced and traded information and goods locally were now surrounded by products they did not produce or even recognize. Products, in effect, had been emptied of their meaning; they had become commodities. And where commodification empties meaning, advertising "fills."[44] Advertisers most often fill products with meanings connected to personal well-being, excitement and success. Ads typically suggest that commodities "offer magical access to a previously closed world of group activities."[45] The act of consumption, it is implied, will reconnect the consumer with the community, or at least *a* community. Not a political community necessarily, but a "community of consumers."[46]

Ownership Consolidation

The economics of centralization combined with the rise of advertising and the public's appetite for newspapers made Pulitzer, Hearst and other major publishers extremely wealthy and powerful men. Hoping to export their models of journalism across the country, they began to purchase national magazines and newspapers in other large cities and to open brand new outlets in others, creating newspaper chains that began to dominate the industry. This was a development that brought "disastrous results to independent newspapers," according to Harold Innis.[47] Roy Howard, a publisher who helped start the prominent Scripps-Howard chain, was quite candid about his desire to attain profits above all else. "We come here as news merchants," he was quoted as saying as he announced the launch of a paper in Detroit. "We are here to sell advertising and sell it at a rate profitable to those who buy it. But first we must produce a newspaper with news appeal that will result in a circulation and make that advertising effective."[48] Innis suggested that in such an environment, a "journalist became one who wrote on the backs of advertisements" rather than a watchdog of the public interest.[49]

Press consolidation led some to question the direction in which the press was moving, namely whether or not the powerful new media moguls could possibly offer reliable news coverage given their increased dependence on corporate advertisers, which some believed would steer them away from stories about corporate corruption, and their overall ability to influence public opinion about issues they covered vigorously. Hearst in particular was known for tampering with his news content so that it would serve both his commercial and political interests. In 1898, he used his *New York Journal* to promote war against Spain after the USS *Maine* blew up near Havana, Cuba. Though there was no evidence to indicate that the Spanish had anything to do with the incident, Hearst published headlines and stories implicating them and called on the government to go to war.[50]

The Progressive Era and Professional Journalism

In response to the blatant sensationalism and irresponsible reporting of yellow journalism and the fear that media owners such as Hearst had become too powerful, a movement to reform the press began. In a sense, there was a fear that the press was too closely aligned with economic and political powers, which were never totally displaced even after the demise of the partisan press. As such, the public sphere was in danger of becoming a zone of society in which elite groups, including the corporate press, would come to dominate public opinion while the greater public was transformed into a passive "mass" audience with limited communicative and hence political freedom.[51]

Criticism of the press corresponded with other efforts to combat the excesses of industrial capitalism—or to eliminate it altogether—that marked the Gilded Age. Often referred to as the Progressive Era, this time of social activism brought to the fore the issue of social responsibility, not only that of the government and corporations but also of the press. In the 1880s and '90s, just as yellow journalism was taking off, the role of the individual journalist, as well as the relationship between reporters, editors and owners, was beginning to be questioned. Media scholars have noted that the word *ethics* was first used in a piece of press criticism in 1889. One year later, the first "code of conduct" for journalists appeared.[52] This indicates that the commercialization of the press along with the specialization of reporters and editors was causing journalism to be viewed as a profession—one that was becoming increasingly influential in public affairs. In order to justify and legitimize this power, "a discourse on journalistic behavior, standards of performance, and professionalism" was in order, according to many press watchers.[53]

At the same time sensationalism and commercialism were being criticized, some members of the press were distinguishing themselves by producing substantive investigative reports that exposed the corruption of political machines, such as New York City's Tammany Hall run by the notorious William Marcy "Boss" Tweed[54] and the greed of corporate robber barons. In doing so, they illustrated the plight of industrial workers, especially child laborers, and poor immigrants in much the same way Horace Greeley did a few decades earlier. Notable "muckraking" journalists of the late 1800s and early 1900s including Ida Tarbell, Jacob Riis (who used photography to document the living conditions of New York's poor), Nellie Bly, and Lincoln Steffens helped create the image of the journalist as a populist crusader who served democracy by providing the public with a ruthless accounting of the activities of the powerful. Many would come to see this as journalism's highest purpose. Indeed, as Stephens puts it, such investigative reporting "has seemed one of our best defenses against Orwellian nightmares."[55]

But would such socially responsible reporting survive in a media environment where barriers to entry were being raised and where centralized newspaper chains were becoming more dependent on corporate advertisers for revenue? The solution to this dilemma adopted by many progressive reformers and publishers was not to break up the emerging press monopolies, support unions for journalists, or finance the development of noncommercial papers that would be operated as public trusts (though these ideas were suggested by some critics). Instead, they decided to launch professional training programs in colleges and universities, programs that would emphasize a journalist's ethical responsibility to serve the community and teach journalists to report the news in an unbiased and "objective" manner so as not to favor one faction or viewpoint over another.[56] Many publishers also committed themselves to building a "wall" between the advertising and editorial departments of their papers in order to keep business decisions from influencing news coverage. (This separation and its application to online journalism is discussed in chapter 9.)

The Journalistic Method

Joseph Pulitzer, who had championed both commercialism and progressive, investigative journalism, helped to frame the mission of the press in the early 20th century. "An able, disinterested, public-spirited press," he wrote in 1904, "with trained intelligence to know the right and courage to do it, can preserve that public virtue without which popular government is a sham and a mockery."[57] This clearly had echoes of James Madison in that Pulitzer focused on the mission of serving the public so as to prevent popular government from becoming, as Madison put it, "a farce or tragedy." It also reinforced the belief that the best way to serve the public was to ignore the economic structure of the press and focus instead on cultivating a cadre of "disinterested" journalists who would gather information and verify facts in order to report the unadorned truth. This premise rested on the assumption that sufficiently trained journalists would be able to put aside personal prejudices and resist internalizing the corporate values of their bosses while gathering and reporting the news. To support this project, Pulitzer, who died in 1911, left instructions in his will to establish a journalism school at Columbia University in New York City and create annual prizes for excellence in investigative reporting.

From Pulitzer's initiative and the support of other philanthropists, the professional journalism movement grew. By the 1940s, an estimated 600 colleges and universities were teaching courses devoted to news reporting and writing.[58] Aspiring journalists were taught to purge their reporting of any bias or ideology; the emphasis was placed on the pursuit of independently verifiable facts about public issues and events that could be gathered through enterprise, observation and investigation.[59] This was heavily influenced by a turn-of-the-century movement to apply scientific methods to social research, business management and public policy. There was a growing belief among educators that the truth about social issues could be obtained if the researcher could position him- or herself outside politics, economics or whatever else was being investigated. As political science professor Darrell West puts it, intellectual reformers believed that "managers both in industry and government should be trained to make professional decisions based on the facts, not on personal beliefs or political connections."[60]

Accordingly, journalists were increasingly being taught to view themselves as existing outside issues and events; they were not to be concerned so much with causes or outcomes but with facts, which when piled one on top of the other would reveal the objective truth. As a result, reporters started writing in the third person instead of using their own voice. Information was shoehorned into the "inverted pyramid" format with its "summary lead," which contained the most pertinent facts (who, what, when, where and how) in the first few sentences of an article.[61] Articles written in this style were more easily packaged and distributed to national and regional audiences than earlier literary forms of journalism that often included detailed descriptions of local people, places and customs.

Lippmann vs. Dewey

Somewhat hidden in the shift toward professionalization and objectivity was a new theory of democracy proposed by some leading public intellectuals. Most notable among them was Walter Lippmann, an astute political thinker and newspaper columnist. In 1922 Lippmann's influential book *Public Opinion*[62] articulated this new social vision and suggested what role journalism should play in achieving it. In short, Lippmann believed that modern, technologically advanced societies had become too complicated to be governed by a fully participating public. The "bewildered herd," as he called the masses, were not capable of determining what was in the best interest of society because they were too easily deceived and given to making decisions based on emotion and superstition rather than rational thinking. The solution to this problem, Lippmann argued, was the cultivation of an elite corps of social scientists and policy experts who would advise elected officials, who he called the "men of action," and disseminate their policy recommendations to the public through groups of well-trained, objective journalists. The public would, in turn, be called upon to ratify the decisions proposed by elite planners or to elect politicians from among the field of candidates preselected by them. Thus the function of journalism was not to stimulate public discussion and debate but to present *reality* as determined by the experts, while citizens would be reduced to the role of passive "spectators of action."[63] Underpinning Lippmann's argument was his belief that the process of liberal democracy—engaging in self-government—was less important than efficiently achieving the desired results of government, that is, the good life.

Taking serious issue with Lippmann was John Dewey, the pre-eminent American philosopher and educational theorist of the period. In a series of lectures, articles and books, notably *The Public and Its Problems*, Dewey challenged Lippmann's views about democracy and the media. Though Dewey shared many of Lippmann's concerns about the complexities of life in industrial society and the limitations of popular government, he did not agree that vesting political and communicative power in an "intellectual aristocracy" was the proper solution. "No government by experts," Dewey wrote in 1927, "in which the masses do not have the chance to inform the experts as to their needs can be anything but an oligarchy managed in the interests of the few."[64] The "few" that Dewey had in mind were the captains of industry, who he saw as the shadowy power behind government. Any "specialized class" of policy experts, Dewey argued, that was not accountable to the public would likely become the commissars for big business.

Instead of Lippmann's model, Dewey advocated a vision of democracy in which the process of self-government was as important, if not more important, than the results. He believed that without freedom, solidarity, a choice of work, and the ability to participate fully in society, individuals could never achieve meaningful lives. Thus the public had to be actively engaged in political, economic and cultural affairs. This required a public sphere through which intellectuals, the public, and elected officials could participate in lateral, rather than hierarchical, communication. As a vital part of this sphere, journalism would fail if it simply tried to inform the public. What it needed to do, suggested Dewey, was open up space for an ongoing dialogue, a cultural conversation consisting of information, signals and stories that would activate inquiry.[65] This would certainly require journalists to be educated and well trained in the art of investigation, but they should not consider themselves to be outside of society—to be neutral observers. Instead, Dewey saw journalists as both social investigators and facilitators of democratic debate. They too had a stake in society, a stake they shared with the rest of the public. Dewey also argued that in order to serve the public better, journalists should have more autonomy over their craft. If they were liberated from commercial pressures, he believed, they might be able to use communication technology in a more artful way. That is, they could creatively explain complex social and scientific themes so that the public had a better understanding of the human consequences of proposed policies.

In the minds of many in politics, business and journalism, Lippmann's "democratic realism" gradually prevailed over Dewey's participatory vision. Reporters continued on a track that emphasized objective analysis and inside connections with policy experts and political leaders rather than two-way communication with the general public. There were exceptions, of course, as populist and radical newspapers, from both the political left and right, vigorously opposed "scientific" government ruled by an intellectual priesthood.[66]

Canons of Journalism

As objective, professional journalism became more established, new professional associations sought to codify editorial standards. One of the first was the American Society of Newspaper Editors (ASNE), which established its "canons of journalism" in 1923. "The primary function of newspapers," the ASNE declared, "is to communicate to the human race what its members do, feel and think." In order to properly tell society what it was doing and thinking, newspapers were urged to adhere to the following guidelines:

Responsibility. The right of a newspaper to attract and hold readers is restricted by nothing but considerations of public welfare.

Freedom of the press. Freedom of the press is to be guarded as a vital right of mankind.

Independence. Freedom from all obligations except that of fidelity to the public interest is vital.

Sincerity, truthfulness, accuracy. Good faith with the reader is the foundation of all journalism worthy of the name.

Impartiality. Sound practice makes clear distinction between reports and expressions of opinion. News reports should be free from opinion or bias of any kind.

Fair play. A newspaper should not publish unofficial charges affecting reputation or moral character without opportunity given to the accused to be heard; right practice demands the giving of such opportunity in all cases of serious accusation outside judicial proceedings.

Decency. A newspaper cannot escape conviction of insincerity if, while professing high moral purpose it supplies incentives to base conduct, such as are to be found in details of crime and vice, publication of which is not demonstrably for the general good.[67]

From the 1920s to the advent of World War II, these standards were promoted across the commercial news industry. When radio emerged as a popular medium, many broadcasters offering news programs formally adopted editorial standards similar to those of the ASNE. This may have been due largely to the belief carried over from the early commercial press that overtly partisan reporting would not attract large enough audiences to satisfy advertisers. But not all radio programmers preferred the objective style. Many allowed populist religious and political leaders to present long speeches that advocated working-class solidarity. While radio's popularity grew rapidly, many newspaper owners worried that the immediacy and intimacy of the medium would draw people away from reading newspapers, especially in the evenings. Studies from the 1930s indicate that the public was giving newspapers low marks for credibility and, indeed, turning to radio for evening news because they saw it as a more participatory medium. Though they might not be able to speak directly with the announcer, audiences could hear another human voice discuss the major issues and events of the day.[68]

Thus while newspapers generally followed Lippmann's communication model, radio appeared to be following Dewey's. This had major implications during the Great Depression. Throughout the 1930s, many major newspapers strongly opposed Franklin Roosevelt's New Deal programs, which were popular with the public. Roosevelt in turn countered press criticism by going on the radio and speaking directly to the public through a series of "fireside chats." In response to the threat from radio, major newspaper owners bought radio stations or invested in national radio networks, which when combined with newspaper chains and wire services resulted in further media consolidation. In addition, the Federal Communications Commission (FCC), created in 1934, sided with commercial broadcasters and the advertising industry in a dispute with educators, unions and religious groups over how best to manage the public's airwaves. The result was the establishment of a market-based, commercial broadcasting system that included only a small number of frequencies allocated for noncommercial, community use.[69]

The rigid standards of objectivity advocated by dominant commercial news outlets were nearly impossible to live up to. Research by noted scholar Paul Lazarsfeld in 1940 indicated that political reporting was still skewed by close relationships between newspaper publishers and leaders of political parties.[70] Beyond the question of party loyalties, some critics echoed some of Dewey's arguments and asserted that the press was poorly serving democracy partly through its corporate ownership structure and insider status, which further alienated the public at a time when they desperately wanted to participate.[71] World War II had something of a unifying affect on the country from which the press and government benefited, though there were still questions about the path democracy and journalism should take. Noted publisher Henry Luce, founder of *Time* magazine, went so far as to establish a commission in 1942 charged with identifying the obligations of a free press in democratic society. Luce was partly motivated by fears that a public that was unhappy with the press might demand government regulations.

The Hutchins Commission

The panel founded by Luce would come to be known as the Hutchins Commission in honor of its chairman, Robert Hutchins, president of the University of Chicago. The group consisted of 12 of the country's leading public intellectuals from the fields of theology, social science, law, history, commerce, literature, and culture.[72] In their deliberations, the commission observed that while the country's population kept increasing, the number of daily papers had declined due to press consolidation. What competition did exist tended to be among sensational tabloids that filled their pages with coverage of celebrities, crime and bizarre events. Commissioners also lamented the lack of substantive reporting on important public issues in the commercial press. They also noted that despite professionalization and the supposed separation of politics and business from the editorial process, many publishers and owners, including Luce, often steered news coverage in directions that were favorable to their political views and financial needs; that many publishers rejected the idea that they had any social responsibility; that coziness with "official" sources and the political establishment often resulted in self-censorship; and that the news was highly segregated—blacks were practically invisible except for crime reports.

Despite these failings, the commission as a whole was not in favor of government press regulations (broadcasting was already regulated due to the scarcity of the radio spectrum). Like James Madison before them, several members worried that self-interested politicians serving a particular party or faction would use regulatory powers to censor opposing views, which would further damage democracy. Some commissioners suggested that the government could provide infrastructure or tax incentives to those who wished to start new papers in cities with only one daily paper. This too was largely rejected. In order to address the narrow range of opinions and racial perspectives in the news, one commissioner proposed that newspapers be treated as "common carriers" to which dissident or underrepresented groups had a right to access and present their views, while another suggested a "right of reply" statute that would allow those whose character or beliefs had been attacked to issue a response. But these ideas ran up against the argument that such measures might prevent journalists from taking on controversial issues and, worse, amount to government control over the editorial process. One plan that did generate broad support was a proposal to start a nonpartisan, nongovernmental organization that would monitor the press for inaccuracies, distortions and bias in addition to promoting "the freedom, the public responsibility and the quality of mass communications."[73] As for the country's journalism schools, commissioners agreed that they should do more to cultivate critical-thinking, socially responsible journalists, though some members were skeptical that this would occur. They argued that close ties to the media industry had turned many journalism schools into narrowly focused vocational programs and that instructors tended to be apologists for, rather than critics of, press failures.

After five years of on-and-off deliberations—interrupted obviously by the war—the commission released a report titled *A Free and Responsible Press*. In it the commissioners stated that a free press was vital to democracy, but they cautioned that such freedom was in danger because the press was underserving the public. The press must, therefore, exercise its freedom in a socially responsible manner or risk losing it. Unless the press radically reformed itself, the report argued, dissatisfied citizens might rebel as Luce feared and urge government to step in and regulate the industry. Thus the report sounded a clear warning:

> If modern society requires great agencies of mass communication, if these concentrations become so powerful that they are a threat to democracy, if democracy cannot solve the problem simply by breaking them up—then those agencies must control themselves or be controlled by government. If they are controlled by government, we lose our chief safeguard against totalitarianism—and at the same time take a long step toward it.[74]

The bulk of the report emphasized the ethical responsibilities of the press, including voluntary acceptance of the principle that they were common carriers of public information and should strive to produce intellectual and artistic content that engaged the audience even if it was not likely to generate instant profits. They also encouraged the press to participate in substantive self-criticism, to raise the competence of reporters and editors, and (for radio news) to clearly distinguish advertisements from news programming. Based on all of its research, the commission concluded that a vital democracy required and a responsible press should deliver news reporting that offered

- a truthful, comprehensive and intelligent account of the day's events in a context that gives them meaning;
- a forum for the exchange of comment and criticism;
- the projection of a representative picture of the constituent groups in the society;
- the presentation and clarification of the goals and values of the society; and
- full access to the day's intelligence.[75]

The emphasis was clearly on things not given primacy in the "canons of journalism," such as the "context" and "meaning" of news, a "representative picture" of society, and social values. In a sense commissioners were urging the press to do some of the things Dewey had suggested while also maintaining a devotion to "truthful" balanced reporting.

Commissioners also had some recommendations for the public and for government. For the public, they suggested the establishment of nonprofit institutions to monitor the news media and to support the production of noncommercial news and cultural programming required by the American people. Additionally, they recommended the creation of academic-professional centers of advanced study, research and publication in the field of communications, and that existing schools of journalism exploit the total resources of their universities so that students would obtain a broad, liberal education in the arts and sciences. For government the commission recommended

- that the constitutional guarantees of the freedom of the press be recognized as including radio and motion pictures;
- that government facilitate new ventures in the communications industry, that it foster the introduction of new techniques, that it maintain competition among large units through the antitrust laws, but that those laws be sparingly used to break up such units, and

that, where concentration is necessary in communications, the government endeavor to see to it that the public gets the benefit of such concentration;
- an alternative to the present remedy for libel, the adoption of legislation by which the injured party might obtain a retraction or a restatement of the facts by the offender or an opportunity to reply;
- the repeal of legislation prohibiting expression in favor of revolutionary changes in our institutions where there is no clear and present danger that violence will result from the expression; and
- that the government, through the media of mass communication, inform the public of the facts with respect to its policies and of the purposes underlying those policies and that, to the extent that private agencies of mass communication are unable or unwilling to supply such media to the government, the government itself may employ media of its own.[76]

Industry reaction to the Hutchins report was mostly negative. Some criticized the report for saying what most press critics already knew and had been writing about for years and for failing to offer any original structural reforms that could address press shortcomings. Henry Luce thought that the report was "uninteresting."[77] Others nailed the Commission for not having any journalists among its elite, intellectual members, while some went so far as to label the members as both fascists and communists.[78] Leaders of several of the country's leading journalism schools also attacked the report for calling on journalists and journalism schools to criticize the press. They contended that respect for the press and cooperation from universities was a better approach.[79]

Though the report was sharply criticized by the press, some of the recommendations of the Hutchins Commission gradually appeared to have some impact. Reporters became more technically competent and more broadly educated in journalism schools that accepted many of the report's principles; independent news councils, journalism reviews and other institutions devoted to media criticism evolved and had some success at getting the industry to look critically at itself; and sensationalism decreased in newspapers over the next few decades, though journalism scholar Stephen Bates points out that it resurfaced later on television news, which, by the 1980s and 1990s, had influenced newspapers to become flashier, more colorful and more sensational.[80] Bates also observes that press consolidation has only gotten worse in recent decades with the emergence of global media conglomerates that control a staggering number of news organizations across media formats. Contrary to the wishes of the Hutchins Commission, successive presidential administrations, Congress and the FCC have actually created a favorable regulatory climate for media mergers and the expansion of global communications conglomerates.[81]

Journalism in the Broadcast Era

Ascendancy of Television

The development of television did not alter the basic regulatory framework for broadcasting established in the 1930s. Many of the same companies that had become dominant in radio, such as NBC and CBS, also launched television operations. Prominent newspaper publishers who had initially been skeptical or fearful of television—as they were with radio—soon realized that the new medium, which brought both words and images into the living room, could be a powerful dissemination tool. Eventually many of them would enter the TV business too. Perhaps most pleased with media policy was the advertising industry, which saw television as a way to transport product images into the intimacy of the home.[82] Prominent industry reports have even referred to television as a "magical, marketing tool."[83]

While television didn't change the regulatory structure, it significantly altered the news consumption habits of Americans and the stylistic structure of news. During television's first decade, the 1950s, the public still received most of its news from daily newspapers. In 1959, for example, 57 percent of those surveyed named newspapers as their primary source of news, 51 percent named television, followed by 34 percent for radio and 8 percent for magazines. Over the next few years, the spread of TV sets into more homes combined with a series of major national and international events (the Nixon/Kennedy debates, the Cuban missile crisis, the assassination of JFK and murder of Lee Harvey Oswald, and the Civil Rights movement), some of which were broadcast live, made television the preferred news medium. In 1963, 55 percent of those surveyed said they got most of their news from TV, 53 percent named newspapers, while 39 percent named radio. Newspapers would never again be the dominant news medium in America. By the mid-1980s, 66 percent of Americans would name TV as their major news source, while only 36 percent named newspapers and 14 percent cited radio.[84] It would be safe to say the television has become the most common household appliance, the most pervasive communications medium. By the mid-1990s, the percentage of American households with TV sets was higher than the percentage that had telephones.[85]

Images and sounds. The ability of the television to make viewers feel as though they are eyewitnesses to major events clearly contributes to the popularity of TV news. The medium plays on the natural instinct to trust moving images, live action and live sounds, more than still words on a page. What people see on TV (especially since the advent of color TV) or hear on TV looks and sounds a lot like what they see or hear with their own eyes and ears. Thus visual scenes and sound bites have become the primary elements around which television news reports are organized. Naturally this diverges from newspaper journalism, which is organized around facts. To be sure, the images and sound bites of TV news do convey facts, but whereas a newspaper reporter can supply a fairly wide range of facts, a TV reporter is typically limited to those facts that can be recorded with a camera or broadcast live. It is often said in television newsrooms that a particular story is a "newspaper story" because it does not consist of an event that will provide "good video and sound." As such, TV reporters are constrained by the technological structure of the medium.

In addition, the commercial imperatives of broadcasting often limit the types of images and sounds that get on TV. The rush to attract audiences with purchasing power, which can then be sold to advertisers, puts a premium on stories about spectacular events or celebrities, preferably a spectacular event featuring celebrities—anything likely to produce good video and sound. In this way, the range of subjects presented on television is narrowed. Media critic Jeffrey Scheuer argues that commercial television "celebrates site, spectacle, and personality" at the expense of complex subjects, such as political and economic problems, and abstractions.[86] Journalism professor Philip Meyer agrees and adds: "Most news coverage is about events because events are cheap and easy to cover. Television's preference for bloody images is not due to cynical manipulation of primitive tastes so much as the fact that it doesn't take a lot in the way of money or brains to chase an ambulance and shoot what it arrives at."[87]

To be sure, ideas, issues and events that don't produce shocking or stimulating images and sounds do get some exposure on news magazine or interview shows, but even these programs tend to emphasize drama or conflict over rational, respectful discourse. On this subject, the late French sociologist Pierre Bourdieu observed that it was nearly impossible to be caught thinking on TV because thinking requires one to be silent for a few moments—or much longer—in order to organize one's thoughts and select the words with which to express them. Just as much as commercial TV abhors a blank screen, it abhors silence.[88] The element of concision also tends to prevent unorthodox ideas from being fully explored. Complicated arguments that attempt to take apart received ideas cannot be blurted out like catchy slogans or clichés, nor can they always fit neatly between commercial breaks. Thus interview shows tend to shy away from guests who slowly build arguments with evidence and logic. As a result of these qualities, media critic Neil Postman suggests that Americans may be the most entertained and least informed (about public affairs) people in the industrial world.[89]

Abandoning the political. This is not to say that television has never produced thoughtful debates and substantive coverage of complicated issues. Initially, TV news was dominated by live coverage of public policy issues—that is, congressional hearings, political conventions and speeches, civil rights marches—or it consisted of lengthy filmed segments on public affairs, documentaries and one-on-one interviews with activists, intellectuals and labor leaders. TV reporters and anchors generally took great care to avoid sensationalism. But as television production became more sophisticated and the commercial demands increased, the cameras began shifting to entertaining or sensational events, people and locations. The emphasis was placed on quick edits, dramatic visuals, graphics and shorter sound bites.

As several well-known studies in the 1990s pointed out, the average length of a sound bite from a political candidate shrank from 43.3 seconds in 1968 to 8.4 seconds in 1992.[90] Paralleling the drop in political discourse on TV has been the explosive growth of paid political advertising, which is a financial windfall for broadcasters. One report indicated that voters looking for campaign news during the height of the 2002 election season were "four times more likely, while watching their top rated local newscast, to see a political ad rather than a political story."[91] Perhaps most distressing, it appears that both the shrinking sound bite and the avalanche of political advertisements have contributed to sharp declines in voter participation among young people.[92]

This is not to suggest that television is solely to blame; it may have only accelerated trends that have been a long time in the making. In fact, voter turnout rates have declined throughout most of the 20th century. Media historian Michael Schudson points out that based on voter participation, the "golden age" of American political culture was the period from 1840 to 1890, when voter turnout rates were as high as 80 percent in some parts of the country.[93] This period, of course, included the penny press, partisan journalism, and the early days of Joseph Pulitzer's muck-raking "new journalism."

Perhaps there is, as some scholars have suggested, a connection between the rise of commercialism and Lippmann-style "objective" reporting and the decline in voter participation. As legal scholar C. Edwin Baker puts it, "the objective style makes politics less accessible, the right answer less clear, the competing parties and candidates less different, and the importance of the choice more difficult to discern."[94] In making similar arguments, other scholars use the word *objective* to mean the practice of assuming a neutral position when reporting the news. As the Hutchins Commission discovered in the 1940s, genuine neutrality was never really the norm; partisanship may have been hidden by the rhetoric of objectivity, but it never disappeared. This being the case, some scholars have argued that hidden partisanship is more harmful and, hence, unethical than overt partisanship because it ultimately deceives—or attempts to deceive—the audience. This is not to suggest that there are no objective truths to be uncovered by reporters—there certainly are and they are supported by evidence. As the noted historian Eric Hobsbawm points out, "Rome defeated and destroyed Carthage in the Punic Wars, not the other way round."[95] It may be quite natural and useful, as Philip Meyer argues, for journalists to have a clearly stated point of view or framework that guides their reporting, so long as they back it up with rigorously tested evidence. The goal, Meyer argues, is "objectivity of method, not objectivity of result."[96]

In addition to the objectivity movement and media consolidation, which has fueled commercialism, others have suggested that television's stimulating visual quality and the ease with which one can use it (passively) has contributed to the withdrawal of citizens from politics and community activities in general. From this perspective, television, videos and other new technologies that efficiently deliver entertainment may be "driving a wedge between our individual interests and our collective interests."[97] In a sense, these theories hold that the news, which more people get from TV than from any other source, has become so divorced from community concerns and the political process that it has contributed to apathy and disengagement among citizens.

Criticism from journalists and the public. It's not just scholars and media critics who lament the decline in community and public affairs journalism on TV and in other media. Journalists themselves have become increasingly critical of their own profession. A report by the Pew Center for the People and the Press released in 2000 indicated that only 37 percent of the national reporters and 35 percent of the local reporters surveyed gave their profession high marks; this was a big drop from a similar survey taken a year earlier. The Pew report also showed that about 25 percent of the journalists surveyed said they "purposely avoided" important stories, while almost as many reporters said they "softened" stories to benefit the interests of their employers. In addition, 41 percent of the reporters surveyed admitted that they have engaged in "either or both" of these practices.[98]

Beginning in the 1980s and continuing through 2002, studies also show that journalism's credibility with the public has declined, though it did briefly go back up in the aftermath of the September 11, 2001, terrorist attacks in the United States. Contrary to the claims of many media consulting firms, the Project for Excellence in Journalism (PEJ), a nonprofit organization that monitors media performance, suggests that citizens do want to see and hear more substantive reporting on politics and other public issues. However, most survey questions about public interest in politics are overly broad, according to the PEJ, which makes them essentially useless. In partnership with the Pew Research Center, the PEJ compared audience responses to questions typically asked by consulting firms and those that were more specific and narrowly focused. The results of the study showed that only 29 percent of the respondents expressed much interest when asked how they felt about political reports in general. "Yet when people were asked whether they'd be interested in 'news reports about what government can do to improve the performance of local schools,' the percentage of 'very interested' jumped to 59 percent."[99]

The Public Journalism Movement

In response to the journalism crisis of recent decades, numerous educators and reform-minded journalists launched a movement designed to reconnect journalism with local communities. Bearing the title "public (or civic) journalism," this movement raised many of the same questions that were debated by Dewey and Lippmann in the 1920s and the Hutchins Commission in the 1940s. A major figure in this movement has been Jay Rosen, a journalism professor at New York University. In the early 1990s, Rosen began arguing that much of what Dewey said in the 1920s was right, especially his assertion that democratic government cannot function without a public that is both informed and empowered to participate in making decisions about how major institutions in society ought to be organized. Rosen's revival of this idea and his argument that journalists have an ethical responsibility to nurture democratic discourse resonated with several newspaper publishers and some TV news directors, who had seen their organizations become highly efficient and professional, but at the same time noticed that the public was becoming more cynical and detached. What could they do to reconnect with readers and viewers, many wondered.

Rosen suggested that journalists could more effectively engage their audiences and help reconstitute the public sphere by striving to (1) address people as citizens, potential participants in public affairs, rather than victims or spectators; (2) help the political community act upon, rather than just learn about, its problems; (3) improve the climate of public discussion, rather than simply watch it deteriorate; and (4) help make public life go well, so that it earns its claim on our attention.[100]

Many of the publishers and TV producers Rosen consulted with began holding community meetings designed to bring people from different economic, political and cultural groups together to discuss ways to increase civic involvement and identify pressing social problems. Based on feedback from these sessions, many publishers decided to adjust their coverage so that more attention and resources were devoted to producing lengthy investigative reports into issues such as education, health care, racism and economic opportunity. In addition to examining social problems with an eye on locating causes, editors and reporters also produced stories that debated possible solutions and featured the efforts of communities and organizations that had success alleviating problems. When it came to crime news, which had become by far the most dominant topic on TV news, reporters shied away from the "if it bleeds, it leads" logic and addressed crime as they did other social problems, probing the community looking for causes and possible solutions. In Charlotte, N.C., the killing of two police officers prompted the city's major newspaper and a TV station to shun sensational coverage in favor of in-depth reports based on community forums where citizens laid out their concerns and proposed solutions. In the field of politics, several papers and TV stations refused to do stories that followed the typical "horse race" angle (who's ahead in the polls) and instead asked the public which issues they wanted the candidates to address. Reporters then took the viewers' concerns and questions straight to the candidates, often causing them to break away from their rehearsed stump speeches and speak about subjects they would rather have ignored.[101] By the late 1990s, Rosen reported that over 300 news organizations from print, TV and radio had participated in public journalism projects.[102]

Critics of public journalism. After its initial growth, the momentum of public journalism was slowed somewhat by a backlash from many prominent national publishers, editors and media critics. Some argued that it forces journalists to play a direct role in shaping the news, which causes them to lose their neutral, independent status. Another complaint was that local and regional coalitions of papers and TV stations participating in public journalism projects tended to "homogenize" the news and actually worked to limit the perspectives presented on TV and in the press.[103] Others argued that public journalism projects were being used primarily as marketing tools. Journalism professor Ron Smith quotes *Washington Post* executive editor Leonard Downie Jr. as saying, "Too much of what's called public journalism appears to be what our promotion department does, only with a different name and a fancy evangelistic fervor." Downie worried that journalists would spend more time worrying about enhancing their image in the community than tracking down leads on important stories.[104] On another level, some scholars have criticized public journalism for ignoring "the structural factors of ownership and advertising" that led to the crisis of journalism in the first place. Communications professor Robert W. McChesney argues that such projects might have more of an impact if they worked to increase the autonomy of journalists.[105]

In response to critics Rosen has argued that "re-engaging people in public life is not the same thing as becoming a partisan interest or advocate for a cause." He also suggests that "it is possible to challenge the community and tell disturbing truths while still supporting a healthier public climate." Finally, he contends, "the press is already an influential actor in politics and civic affairs, not a bystander." Thus it is actively engaged in society whether journalists acknowledge it or not. Ultimately, Rosen believes that journalism "can learn to use its influence on behalf of a strengthened democracy."[106] Author and journalist James Fallows has also defended public journalism from its critics. In particular he addresses the charge that civic engagement subverts independence and neutrality. "One of public journalism's basic claims," he writes, "is that journalists should stop kidding themselves about their ability to remain detached from and objective about public life." Fallows contrasts journalists with scientists and observes that unlike scientists journalists "inescapably change the reality of whatever they are observing by whether and how they choose to write about it."[107]

The Internet Age

By the early 1990s a new era had begun—the era of the Internet. The Internet demolished the existing barriers of entry to publishing and as a result has led to entirely new forms of communication. This age raised the question of who should be called a journalist when anyone can disseminate his or her news and views to the world. This is the topic of the following chapter.

Summary

What we know today as professional journalism did not begin until late in the 19th century, although there were earlier versions of a kind of press.

During colonial times, pamphleteers and public thinkers circulated materials that influenced the body politic, which gave voice to dissenters during the American Revolution. Much of the press of the 1800s fell under a category called the "partisan press," as these were newspapers aligned with one of the two major political parties, the Federalists and the Republicans. During the 1800s, as the numbers of newspapers grew they became more a commercial enterprise, were supported by advertising, became cheaper (the "penny press") and reached more of a mass audience. An example of a newspaper that exemplified the emerging populist paper model was the *New York Sun*. Other major papers of the 1800s included the *New York Herald,* and the *New York Tribune.*

The *Tribune* distinguished itself from the *Sun* by reporting less gossip and doing more investigations into social issues. Competing with this model was the sensational or "yellow" journalism press, best illustrated by William Randolph Hearst's newspapers. Another emerging model during this time was an "information" model, which was practiced by *The New York Times*. The *Times* aimed at professionals and strived to provide an objective accounting of the factual matters of the day.

During the Progressive Era of the 1880s and 1890s, as the spread of a sensationalist press and the political influence of its publishers increased, a movement to reform the press grew in reaction. This reform movement led to more investigative reporting of current social ails and a new focus on ethics and codes of conduct.

By the early 20th century, there was a movement to create a more independent, objective and public-spirited press, as a way to better inform the public, and make the population effective citizens. This marked the beginning of modern professional journalism. During this period Walter Lippmann and John Dewey articulated two competing theories on the appropriate role of journalism in society. Lippmann believed the public was not rationally capable of making good decisions and so believed in the value of creating a cadre of knowledgeable authorities and a core of elites who would present expert views to the masses. Dewey, in direct contrast, did not believe in the merits of a top down form of information dissemination but in a participatory model where the public would dialogue to come to its own decisions. In Dewey's view, a key role of journalism was to facilitate this kind of communal conversation. It's widely thought that Lippmann's view became the dominant model for the media, though certain exceptions followed a more Dewey-like approach.

As the new professional journalism became established, journalism associations such as the ASNE created ethical standards and codes. The 20th century also brought about the advent of radio, which followed more a populist Dewey model. Television emerged and took over a great portion of the public's news consumption. This new medium was subject to much criticism over its abilities to provide rational and thoughtful discourse to viewers.

By World War II, there were increasing concerns and worries about the effect of the media on public discourse. These concerns led to the formation of an investigative panel, the Hutchins Commission. The commission issued an influential report on the media, *A Free and Responsible Press,* that cautioned that the press was falling short on several grounds, including allowing publishers to have too much influence over the news.

By the 1980s, coinciding with the influence of television and a decline in serious reporting, the media began suffering another loss of credibility among the public. One of the responses to this crisis of confidence was a movement called public journalism, which focused on the need for the media to recommit itself to doing a better job in informing the public and assisting them in making important decisions. Proponents felt that reporters should assist their community in facilitating a community conversation about their social problems and actually help the public come to solutions, not just report on the problems. Some in the media criticized public journalism as moving journalists into too much of a role of advocacy and engaging in public-relations- type activities.

A new age for journalism began in the early 1990s with the Internet. The Internet eliminated the traditional high barriers of entry to publishing, and raised the question of who actually should be called a journalist in an era when anyone with access to a PC connected to the Internet can disseminate news and opinions worldwide.

Critical Thinking Questions

- In what ways can the media assist people to do a better job in performing the democratic processes of self-governance?
- Do you think the public is better served by a partisan press or the more current objective press model, and why?
- How did the press differ before it became a profession? Does online journalism resemble the kind of journalism that preceded the emergence of a professional press?
- In what ways does online journalism provide a public discourse differently than print or broadcast journalism?
- Does online journalism embody more the model of Dewey or of Lippmann? Why?

Key Terms

Consolidation
Hutchins Commission
New journalism
Objective journalism
Partisan press
Penny press
Public journalism
Public sphere

Recommended Resources

Books

Bagdikian, Ben H. *The Media Monopoly,* 6th ed. (Boston: Beacon Press, 2000).
Barnouw, Erik. *A History of Broadcasting in the United States,* 3 vols. (New York: Oxford University Press, 1966–1970).
Bates, Stephen. *Realigning Journalism with Democracy: The Hutchins Commission, Its Times, and Ours* (Washington, DC: The Annenberg Washington Program in Communications Policy Studies of Northwestern University, 1995).
Carey, James W. *Communication as Culture* (New York: Routledge, 1992), especially chaps. 3 and 6.
Dewey, John. *The Public and Its Problems* (Athens: University of Ohio Press, 1983). Originally published in 1927.
Fallows, James. *Breaking the News: How the Media Undermine American Democracy* (New York: Vintage, 1997—paperback).
Herman, Edward S., and Noam Chomsky. *Manufacturing Consent: The Political Economy of Mass Media* (New York: Pantheon, 1988). See also the 2002 edition with a new introduction, also by Pantheon.
Kovach, Bill, and Tom Rosenstiel. *The Elements of Journalism: What Newspeople Should Know and the Public Should Expect* (New York: Crown, 2001).
Lippmann, Walter. *Public Opinion* (New York: The Free Press, 1965). Originally published in 1922.
Schudson, Michael. *Discovering the News: A Social History of American Newspaper* (New York: Basic Books, 1978).
Serrin, William, and Judith Serrin (eds.). *Muckraking: An Anthology of the Journalism that Changed* America (New York: New Press, 2002).
Stevens, Mitchell. *A History of News: From the Drum to the Satellite* (New York: Viking, 1988).

Associations and Institutes

Pew Center for Civic Journalism. Web site: www.pewcenter.org/.
Project for Excellence in Journalism. Web site: www.journalism.org.

Universities

U.C. Berkeley Graduate School of Journalism/Resources: Ethics and the New Media. Web site: http://journalism.berkeley.edu/resources/newmediaethics .html.

CHAPTER 3

New Challenges for Journalism: Who Is a Journalist in the Internet Era?

Chapter Goals:

- Identify practical and legal matters as to what is at stake in defining a journalist.
- Surface differing points of view as to what should count as journalism.
- Describe the concept of open journalism.
- Examine Weblogs' relationship to journalism.

In an age when anyone with a computer and a modem can, theoretically, gather and disseminate information globally, the question of *who counts as a journalist* is frequently being asked in a number of venues —both physical (traditional newsrooms, college classrooms and courtrooms) and virtual (Web sites, e-mail lists and chat rooms). And what does it mean when someone without specialized journalism training, news experience, or even a particular interest in the craft's values or history can now easily step into a social role that so clearly impacts and shapes the way a democratic society functions? It is difficult to surface a conclusive answer because so much depends on each respondent's point of view.

Without exaggerating too much, it is safe to say that nearly all parties with an interest in the question—individual journalists, judges and lawyers, media corporations, professional associations, journalism unions, media activists, educators and citizens—offer different answers. Some say only those working in traditional media count, and some argue that those with a college degree and/or substantial professional experience qualify, while others suggest that it depends not on the medium or employment status but rather on the intent of the individual—that is, whether or not they are collecting information with the purpose of disseminating it to the public.

Others resist answering the question at all, arguing that journalists shouldn't be defined by any single standard or narrow set of standards. Journalism is a craft, they say, that can be practiced by many people regardless of medium, employment status or intent. Ultimately, they contend, it's the work that matters—let the readers and viewers decide who is or isn't a journalist.

Despite this important argument, legal matters—for example, state-mandated press protections and administrative rules such as those concerning access to government press galleries or sports and entertainment venues—often call for some sort of definition to be established and enforced. As one might expect, the proliferation of online news services makes the access issue difficult to resolve—though the trend is moving in favor of granting press passes to Web sites that produce a substantial amount of original news reporting.

There are also those who maintain that the news media's credibility is further diminished when there is confusion, both within the media and among the public, about what distinguishes a journalist from a nonjournalist. From this view, being called a journalist reassures the public that one is trustworthy; that certain standards are being followed in the gathering and reporting of the news. It is necessary, in the opinion of some journalists, to start a dialogue aimed at redefining their craft. Those who approach the problem this way believe it might be possible to reach a broad consensus among opposing factions as to the higher purpose of journalism and the ethical values that contribute to good journalism, regardless of the medium being used. We will examine the attempt to develop an ethics-based theory of journalism later. But first, it would be helpful to review some of the legal questions involved in deciding who is or isn't a journalist.

Legal Questions

The emergence of the Web as a flexible, inexpensive publishing platform has dramatically lowered communication barriers that historically prevented many people from gathering and publishing news, ideas and opinions. Now, millions of aspiring reporters, cultural critics, media activists and political pundits can publish their own work and make the claim that they are journalists. Though this might upset some experienced, professional journalists who dispute their claim, ordinarily it would not be cause for any legal concern. After all, journalists are not required to carry a license in the United States and the First Amendment is supposed to protect free expression for all citizens.

However, in certain situations, making the claim that one is a journalist has important legal implications. Thirty states have "shield laws" that permit journalists to protect the identity of confidential sources. If a reporter in one of these states tries to invoke the privilege, the courts may have to decide whether or not that reporter qualifies as a journalist. Not surprisingly, the explosion in the number of Web sites providing news adds a new wrinkle. To make matters even more complicated, the courts are very reluctant to get into the business of deciding who counts as a journalist. To many judges—and journalists—doing this would be tantamount to the *licensing* of journalists, which would then raise the specter of government press restrictions and censorship. Indeed, as the Supreme Court wrote in 1972, defining "categories" of journalists who qualify for certain privileges is "a questionable procedure in light of the traditional doctrine that liberty of the press is the right of the lonely pamphleteer . . . just as much as of the large, metropolitan publisher."[1] Despite their reluctance, however, the justices also indicated that the very presence of shield laws would force the courts to decide who is or isn't a journalist.

An article in the May 2002 issue of *Quill*, a journal published by the Society for Professional Journalists, pointed out that many shield laws were written before the Web; therefore, they only refer to professional print or broadcast journalists.[2] Under a literal interpretation of such laws, a writer whose work appears only on the Web might not qualify. However, in 1998, a federal circuit court deciding a case that didn't specifically involve the Internet appeared to create an opening for online writers to invoke journalistic privilege. Drawing heavily on a prominent 1987 case,[3] the judges first supported the argument that the First Amendment was designed to protect the journalistic process rather than individual journalists or the press as an institution. Then they defined the key elements that make up that process: "We hold that individuals are journalists when engaged in investigative reporting, gathering news, and have the intent at the beginning of the newsgathering process to disseminate this information to the public."[4] After issuing their definition, the judges briefly referred to a 1938 case in which the Supreme Court argued, "it makes no difference whether the intended manner of dissemination was by newspaper, magazine, book, public or private broadcast or handbill because the press, in its historic connotation comprehends every sort of publication which affords a vehicle of information and opinion."[5] In light of this, the judges reaffirmed their position that a proper test for determining who is or isn't a journalist should be based on the "intent behind the newsgathering process rather than the mode of dissemination."[6] It appears, then, that Web-based writers who meet such a test could be recognized as journalists by the courts.

It is worth noting, again, that the court was not dealing specifically with the question of whether or not online publishers qualify for shield law protection. Jane Kirtley, a media law and ethics professor at the University of Minnesota, believes that most courts dealing with shield laws would not consider amateur Web writers to be journalists even if their original intent met the appeals court's standard. "If anybody who has a computer and a modem says, 'I'm a journalist,' the courts aren't going to buy it."[7] Ronald Goldfarb, an author and attorney, sounds a similar note when he suggests that intent alone should not determine who is or isn't a journalist. Instead, he argues that a journalist is "somebody who has been judged by the marketplace by being published in a legitimate medium."[8] To Goldfarb, the Web, for the most part, isn't "legitimate" because articles can be placed online without any scrutiny, which, he believes, is a necessary part of the publication process.

The views of Kirtley and Goldfarb differ from that of Lucy Dalglish, a journalist and attorney who serves as executive director of the Reporters Committee for the Freedom of the Press (RCFP), an organization that counsels journalists on a wide range of legal issues. Dalglish resists narrow definitions for fear that they would lead to the licensing of journalists and, subsequently, the erosion of First Amendment protections. If journalists must be defined at all, she prefers the guidelines issued by the federal appeals court judges in the Madden case: A journalist is "someone who is collecting information with the purpose of disseminating it to the public."[9]

It's worth noting that Kirtley, Goldfarb and Dalglish have all worked as professional writers or journalists and as attorneys who counsel writers and journalists—Kirtley was even executive director of the RCFP before Dalglish—yet their views diverge on the question of who is or isn't a journalist. That being the case, and considering the reluctance of courts to wade into such turbulent waters, it's not unreasonable to expect vague or inconsistent guidance in the future.[10]

Ethics and Standards: Defining Good Journalism

While some journalists, their attorneys and the courts continue to wrestle with legal guidelines, other journalists are trying to define their craft by identifying the core values and ethical standards that support good journalism. But here again, there is disagreement as to what those values and standards are, and whether or not they can be translated across media platforms. Without generalizing or simplifying too much, it appears there are two contrasting theories of journalism that are shaping the disagreement. One consists of established standards and practices that emanate from print and broadcast journalism and the belief that journalism has a social responsibility to inform citizens and nurture democracy, while the other is informed by suspicion of centrally managed, traditional media conglomerates and a belief, inspired by the open architecture of the Internet and the flexibility of Web publishing, that citizens can participate in democracy by creating their own journalism. We can call the first theory "traditional journalism," for lack of a better term, and the second theory "open or participatory journalism."

Traditional Journalism

Traditionalists often make the argument that journalists perform a necessary "gatekeeping," or filtering, function in society. As events happen, journalists gather facts, organize them, edit out irrelevant or redundant information—the "noise" in the channel—and deliver what's left to consumers. Or as Dan Okrent, editor-at-large for Time, Inc., describes it, journalists "take the world and put it into a funnel . . . we get rid of the bad stuff and presumably give the reader or the viewer the good stuff."[11]. This news "funnel" is akin to a production process in which journalists subject information to an established set of standards, adapted from the objective model of journalism, before publication. Jim Godbold, a newspaper editor in Eugene, Ore., says these standards include "truthfulness, verifiability, fairness, and completeness."[12]

In addition to this quality-control task, traditionalists assert, journalists also help the audience make sense of information through analysis and interpretation. Thus, they are more than mere fact checkers as John Granatino, an executive with the Belo media company, explained in an e-mail discussion with other journalists, "Good journalism brings perspective and reason to situations where emotion and ignorance can lead poorly informed people to jump to unwarranted conclusions."[13] The journalist-as-gatekeeper, then, delivers accurate news with the voice of reason when complicated events and information threaten to overwhelm citizens. This is essentially a modern adaptation of Walter Lippmann's model of journalism.

Traditional Journalism in a "Mixed Media Culture"

Traditionalists fear that the rapid emergence of Web publishing both by amateurs and professionals—and the simultaneous expansion of cable TV news—has helped spawn a media culture that threatens the credibility of traditional journalism and undermines the gatekeeping function. According to Bill Kovach and Tom Rosenstiel, veteran journalists who authored the book *Warp Speed: America in the Age of Mixed Media Culture*, a "never ending news cycle" has shifted the focus of journalism from "ferreting out the truth" to publishing unverified allegations followed by counterallegations. In addition, the sheer number of media outlets publishing news and information "diminishes the authority of any one outlet to play a gatekeeper role over the information it publishes."[14] While Kovach and Rosenstiel acknowledge that upstarts who don't adhere to established standards and practices can open up journalism to innovation and new voices, they also contend that a "wider range of standards" has the overall effect of driving out the higher standards. As a result, news organizations are "caught between trying to gather the information for citizens and interpreting what others have delivered ahead of them."[15] On another level, Kovach and Rosenstiel note that the decentralized, open architecture of the Internet dramatically alters the traditional flow of information. As they put it: "The pipeline goes straight to the citizen", giving those so inclined the ability to access the raw material of news, which, in turn, puts even more pressure on reporters to stay ahead of the feed by publishing unverified information or interpreting that which has already been published. In the end, the torrent of unfiltered information threatens to overwhelm both journalist and citizen.

In sum, many traditionalists, Kovach and Rosenstiel prominent among them, assert that journalism has been "in a state of disorientation" brought on in large part by "rapid technological change" and economic pressures.[16] With attention focused on technology and business interests, journalists have forgotten what distinguished their craft from other forms of communication. The prescription, according to Kovach and Rosenstiel, is to recover a sense of what it means to be a journalist —to redefine the craft, taking into account ethical standards, technological change and the needs of the public.

In 1997, hoping to create a modern, more inclusive version of the Hutchins Commission, Kovach and Rosenstiel formed the Committee of Concerned Journalists (CCJ), a group comprised of hundreds of journalists and educators from across media platforms. By 2000, after conducting dozens of public forums and gathering empirical research, they released a "Statement of Shared Purpose" that declared, "The central purpose of journalism is to provide citizens with accurate and reliable information they need to function in a free society." The statement then moved on to a list of nine principles that, in Kovach and Rosenstiel's words, "have ebbed and flowed over time, but they have always in some manner been evident" in good journalism, regardless of the medium.

1. Journalism's first obligation is to the truth.
2. Its first loyalty is to citizens.
3. Its essence is a discipline of verification.
4. Its practitioners must maintain an independence from those they cover.
5. It must serve as an independent monitor of power.
6. It must provide a forum for public criticism and compromise.
7. It must strive to make the significant interesting and relevant.
8. It must keep the news comprehensive and proportional.
9. Its practitioners must be allowed to exercise their personal conscience.[17]

Kovach and Rosenstiel contend that newsgathering and reporting on the Web, or any other platform, can be measured against these principles. They acknowledge that reporters won't meet each standard on every story, but those who succeed more often than they fail are worthy of being called journalists.[18] The two journalists have also conducted numerous workshops in newsrooms and schools across the United States, advocating the nine elements as a model for good journalism. In 2001, the SPJ honored Kovach and Rosenstiel for their research in journalism.

Structural Criticism of the Media

From a more critical perspective, some communications scholars argue that Kovach and Rosenstiel's effort to focus on ethics and the role of the individual journalist draws attention away from the business executives who employ them and the structural factors that influence their work. Robert McChesney, a communications professor at the University of Illinois, argues that giant telecommunications companies with large holdings in journalism—and often in other economic sectors such as professional sports, movies, music and theme parks—constantly pressure journalists, both online and offline, to churn out trivial content that can be cheaply repackaged and reused[19] by other company affiliates. The news media thus operates more like an information and entertainment service within a larger media combine rather than an independent journalistic enterprise. What the craft needs most, McChesney contends, is not a new definition or a new list of principles but liberation from centralized conglomerates driven by the logic of the market rather than public service. Many holding this view argue that no matter what new technologies emerge, they are not likely to foster democratic communication unless the corporate, commercial media model is transformed into a community-based, noncommercial system that is more directly accountable to citizens. As McChesney writes in a 2002 article, "ultimately those who own, and hire, and fire, and set budgets determine the values of the medium."[20]

Open Journalism

While many advocates of open journalism respect the principles articulated by Kovach and Rosenstiel, they resist the attempt to define journalists by translating broad principles across media platforms as well as the economic centralization of contemporary mass media. Jon Katz, a media critic and contributor to the popular technology-based Web site Slashdot, argues that "online, there is no workable definition of what a journalist is . . . Anybody who sees him or herself as a journalist becomes one, which is the way it ought to be."[21]

In taking this position, Katz is suggesting that the architecture of a medium will play a part in shaping the journalistic process, perhaps more so than economic structures. Print and broadcast media, for example, are essentially "closed" communication systems, meaning that information typically flows one way, from news providers to audiences, with little or no mechanism for feedback—it is, essentially, a monologue. The Internet, on the other hand, is an "open" communication system through which information can flow back and forth between news provider and user or between individual users. Thus Katz and others suggest that the Internet helps to democratize communication, offering many points of entry for all who wish to participate in political or cultural conversations.

Open Journalism and Editing

Robin Miller, a former print journalist now writing for Slashdot, also contends that the Internet, as it's currently designed, is "biased" toward sharing stories and engaging in dialogue.[22] Given this architecture, Katz and Miller both suggest that journalism on the Web should be an open, interactive affair in which users participate in gathering and disseminating information as well as correcting the mistakes of others. On this last point, they push the idea that the editing process, which happens behind the scenes before publication in traditional journalism, can be done continuously, even after publication, in full public view on the Web.

This philosophy has connections to the "open source" software movement to which Slashdot (see Figure 3.1) is directly linked (the site is operated by the Open Source Development Network and was built with open source software). This movement is based on the principle that the best software is developed through open collaboration—among users who have access to the "source code" of the software—rather than closed, proprietary arrangements, which block users from seeing the code. Open source developers can alter or customize features on existing software so long as they agree to make their innovations available to other developers.

In addition to Slashdot, several other popular sites, including Kuro5hin (pronounced "corrosion"; see Figure 3.2) and the Independent Media Center (Indymedia; see Figure 3.3) network, advocate open journalism. Though the procedures for submitting stories may differ, each of these sites allows users to post their own news updates and comment on news items posted by other users.

Open Journalism and Accuracy

Many traditional journalists wince at such loose editorial standards, suggesting that they contribute to the spread of unsubstantiated rumor

Figure 3.1 Slashdot (http://slashdot.org):
- Created in 1997 by Rob Malda, now owned by the Open Source
  Development Network (OSDN)
- Focuses on technology and software development
- A team of editors screen incoming submissions
- 400 volunteer "moderators" rate reader comments
- Anonymous submissions accepted
- Registered users can comment on articles
- Registration is free

and gossip already so common on the Web. In response, Miller argues
that, based on his experience, opening the editorial process actually im-
proves accuracy. With more eyes on the content, Miller claims that mistakes are caught and corrected faster than they might be on a Web site
with limited interactivity and a skeleton editorial staff. Rusty Foster, the
creator of Kuro5hin, agrees that "even if something does get published
with less-than-pristine sources, anyone can comment, and quickly confirm or disprove assertions made in the article."[23] It's the instant feedback loop, according to Miller, that forces writers to stay alert, which, in
turn, fosters better reporting.

As one example of open journalism's contribution to accuracy, Katz
and several other media critics have pointed to a story that Slashdot

Figure 3.2 Kuro5hin (www.kuro5hin.org):
- Created in 1999 by Rusty Foster, who is still the main proprietor/owner
- Focuses on technology, culture and politics
- Administered by Foster and eight technical/editorial volunteers
- Only registered users can submit articles and comments
- Registered users decide which stories are posted on the front page
- Registration is free

readers helped edit for *Jane's Intelligence Review*, a publication that focuses on defense issues (see Figure 3.4). The story dealt with cyberterrorism and was scheduled to run in an upcoming issue of *Jane's,* but the magazine's editor had doubts about the accuracy of some of the technical information and, lacking the expertise to verify it, he passed the story along to Slashdot's tech-savvy readers and asked them for feedback. Within a few hours, the story had attracted hundreds of comments and corrections, which led the editor to scrap the article and write an entirely new piece using much of the commentary from Slashdot users.

A writer for Salon.com suggested that the episode marked a "major step forward" for online news. He also closed his article with these words: "There's an immense amount of expertise out there on the Net—sites like Slashdot are pioneering new territory as they facilitate access to that knowledge, to the great and lasting benefit of all of us."[24]

Figure 3.3 Independent Media Center Network (www.indymedia.org):
- Created in 1999 by independent journalists and political activists
- Focuses on political/economic/environmental activism
- Consists of over 90 Web sites in over 30 countries as well as numerous print and radio projects
- Staffed by volunteer editors
- Anyone can publish articles or comments to the "Newswire"
- Editors maintain the "Features" section
- No registration required

In sum, supporters of open journalism feel that they are putting the Web to its proper use. To be sure, open journalism can be a contentious and untidy process, often marked by insults and trash-talk (called "flames") that turns away many potential contributors and leads to the discourse being controlled by domineering personalities or technical elites. What is promoted as democratic discourse can quickly become monologic unless some basic ground rules—short of outright censorship —for respectful communication are applied. Some critics doubt that such standards can be achieved in a disembodied environment such as cyberspace.[25] But supporters also argue that traditional journalism is far more prone to elite control, though its processes are often kept out of sight—print and broadcast media tend to keep their audiences in the dark

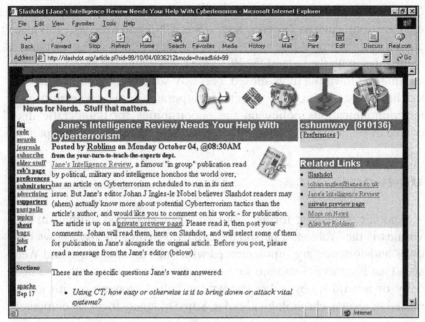

Figure 3.4 October 4, 1999: Slashdot readers are asked to comment on a "cyberterrorism" article being prepared by *Jane's Intelligence Review*.

about the "hows and whys" of newsgathering and reporting. By contrast, "collaborative media puts that process right out in the open," according to Foster. "At least if you find the result inaccurate or unconvincing," he argues, "you can both see how it got that way, and help fix it."[26]

For Katz, this is a refreshing alternative to journalism that consists of "handfuls of editors closeted in offices dictating agendas."[27] Foster also appreciates the sense of liberation that comes from doing open journalism: "We've taken the priestly power of public-opinion-making away from the sanctified towers of The Media, and put it back in the hands of everyone."[28]

Many critical communications scholars argue that it is unrealistic to suggest that Internet populists are about to displace traditional media outlets and their gatekeepers just yet, if ever. They point to evidence showing that not "everyone" is online (about 66 percent of adults in the United States use the Internet, while the percentage in some developing countries is near zero) and of those who are, the number visiting and contributing to open journalism Web sites is a small fraction of those who get their news from major commercial portals, such as America

Online (AOL) and Yahoo!, or Web sites affiliated with traditional news organizations, such as *The New York Times*, CNN and MSNBC. It is also argued that dominant Internet service providers (ISPs) and their parent companies, with the help of the FCC, are redesigning the architecture of the Internet toward a more closed, one-way structure, much like that of the broadcast media. The effect of corporate media ownership and telecommunications policy on the opportunities for democratic discourse through the Internet are discussed more thoroughly in chapter 9.

Weblogs

Another publishing format embraced by many open journalism advocates is the Weblog, a personal journal consisting of short news updates and commentary supplemented with links to other blogs and Web sites (see Figures 3.5 and 3.6 for examples). Blogs are usually updated daily or several times a day (by writers called bloggers), with items posted in reverse chronological order. Some bloggers focus on computer

Figure 3.5 Dave Winer's popular blog, Scripting News.

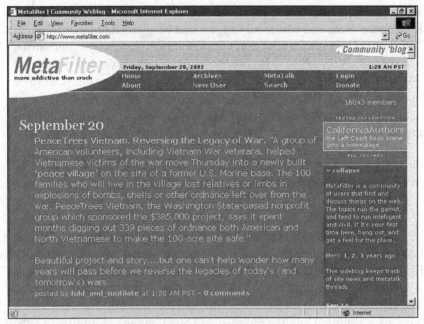

Figure 3.6 A group Weblog called MetaFilter.

technology, politics, or the news media, while others construct personal diaries, publish original poetry, or upload multimedia files—it all depends on the blogger.

Though there are some group Weblogs, which function like scaled-down versions of Slashdot or Kuro5hin, most Weblogs are "edited" as the individual writer updates his or her own commentary. Often the updates are driven by new discoveries made while surfing other blogs or through e-mail comments from other bloggers. As a result, there appears to be a kind of shared editing system at work in what some avid bloggers have termed the "blogosphere." Scott Rosenberg, managing editor of Salon.com offers a perceptive description: "The editorial process of the blogs takes place *between and among* bloggers, in public, in real time, with fully annotated cross-links."[29] In this way, a story grows as it is passed from blog to blog, with each reader contributing something new or challenging some fact they believe to be incorrect. "This carries pluses and minuses," according to Rosenberg, "at worst, it creates a lot of excess verbiage that only the most fanatically interested reader would want to wade through. At best, it creates a dramatic and dynamic exchange of information and ideas."[30]

Bloggers' exchanges often consist of criticism of traditional media. Rosenberg, Katz and Miller all suggest that bloggers and other frequent Web writers, many of whom don't consider themselves to be journalists, have become very adept at fact checking news reports on newspaper and TV Web sites, instantly publicizing mistakes or what they consider to be biased reporting. Rosenberg suggests this is due to "free-floating anger at the professional media's penchant for making mistakes and not owning up to them."[31]

Blogs and Traditional Journalism

As blogs have become more popular—due largely to improvements in self-publishing software—many traditional journalists have begun tapping them for story ideas and opinions, while others have begun publishing their own blogs, taking on subjects they can't approach in their day jobs. Several Web sites connected to traditional media, such as MSNBC.com, the *Providence Journal*, and the *San Jose Mercury News*, have even ventured into the blogosphere by allowing columnists to publish their own blogs on the company's sites. Thus, while many in print and broadcast journalism still look suspiciously at blogs, or don't look at them at all, others are beginning to see them as both liberating forms of communication and valuable newsgathering resources.

Which brings up an interesting question: If traditional journalists don't want to admit bloggers to their ranks and most bloggers don't even consider themselves to be journalists, yet they increasingly contribute to the content of traditional news media, what should they be called? Joe Clark, a Toronto-based blogger and free-lance print journalist, suggests that we might think of them as news *sources* or perhaps as "witnesses" in the tradition of religious speakers and writers who presented first-person accounts of spiritual experiences. In the context of the blogs, witnessing could be construed as a "need" or "obligation" on the part of the blogger to give a firsthand account of what he or she has seen or experienced "so that a form of edification, reassurance, or catharsis can take place" among readers.[32]

On another level, Ron Rosenbaum, a columnist for *The New York Observer,* argues that first-person narratives—an important feature of journalism and nonfiction literature in the 18th and 19th century that was pushed aside in favor of the third-person, "objective" reporting style in

the 20th century—are in need of recovery. Though not referring specifically to Weblogs, Rosenbaum echoes thoughts expressed by ardent
bloggers. The well-written personal account, he contends, is still valuable because it doesn't pretend to be the absolute truth; rather it "lets the
reader into the [thinking and writing] process and tells them: 'This is
what I took into account; this is how I arrived at my perspective; these
are my doubts and hesitations. Take it for what it is'."[33]

Blogs appeared to work on both levels in the wake of the terrorist attacks in the United States on September 11, 2001, as they were quickly
filled with vivid, street-level accounts of the tragedy and reaction from
concerned people around the world (see Figures 3.7 and 3.8). In addition to "witnessing" in the manner Clark describes, bloggers were credited with providing a reliable stream of news and commentary that often
filled gaps left by traditional media. Paul Andrews, a technology columnist for *The Seattle Times,* offered these words of praise in an article
about Weblogs in the *Online Journalism Review:* "While TV stations
replayed ad nauseum footage of the plane colliding with the tower—and
while most newspapers were still running sketchy wire reports—

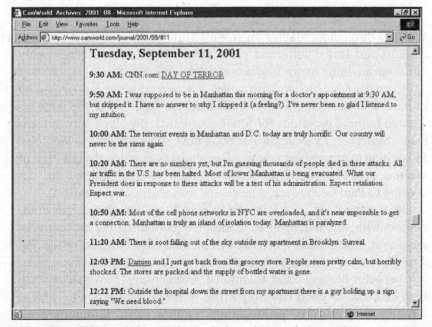

Figure 3.7 Weblog of New York resident on September 11, 2001.

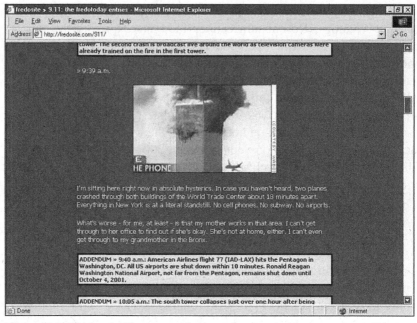

Figure 3.8 Weblog with news updates and commentary.

Weblogs throughout Manhattan provided raw feeds from street level . . . Able to post text, photos and video almost immediately, blogs easily outshone anything major media could provide."[34]

No matter what bloggers are called, it appears that they are beginning to have a substantial effect on journalism. In fact, one of the most respected training programs for future journalists, the University of California at Berkeley's journalism school, now offers a course in Weblogs. In addition to creating a class Weblog, students examine the question of whether or not blogs are "a sensible medium for doing journalism," according to Paul Grabowicz, a professor who teaches the course.[35]

Other journalists are beginning to view blogs not so much as competitors with, or challengers to, traditional journalism, but as a complement to it. Dan Gillmor, technology columnist and blogger for the *San Jose Mercury News,* suggests that there may be a natural division of labor between bloggers and traditional journalists, which makes them both essential to good news reporting. Whereas bloggers are adept at ferreting out "untruths" and locating "a zillion facts," Gillmor argues, traditional news organizations are better at "solid investigative journalism, the kind that takes deep pockets and lots of time."[36] Rosenberg agrees and offers an apt description of the essence of the relationship, "Increasingly, in fact, the Internet is turning it into a symbiotic ecosystem—in which the different parts feed off one another and the whole thing grows."[37]

Ethical Obligations of Bloggers

Whether we choose to call bloggers journalists or something else, the question arises as to whether these individuals should be bound to the same ethical standards as traditional journalists. Can these freewheeling mavericks also follow an existing professional media code? For instance, do bloggers need to check their work for accuracy? How should they handle rumors and e-mails from other bloggers? Must bloggers try to be fair? And do journalists who operate a blog on the side venture into ethically hazardous territory? (See Sidebar for a discussion of this question.)

Sidebar

The Risks of Journalists Who Blog

Not only do reporters use bloggers as sources for interesting grassroots leads but many have also become bloggers themselves. A handful of reporters turned blogger are like journalist Andrew Sullivan, already prominent before beginning their blog. Others are simply beat writers or columnists for a local paper who wanted to branch out and be able to voice their own thoughts, unfettered by editors or other constraints.

But a reporter who starts his or her own blog may be entering into ethically murky territory and will need to tread carefully. This is especially true if the blog is an "unofficial" one, not sanctioned by the paper where the reporter has his or her "day job." In at least one case, in fact, a reporter lost his job because of his moonlighting Weblog activities.

The case involved a reporter from the *Houston Chronicle* by the name of Steve Olafson. Olafson had begun his own Weblog and operated under the pseudonym of "Banjo Jones." Olafson's site engaged in humor and political satire, often poking fun at some of the hometown politicians as well as the local media. But Olafson was covering the same officials in the *Chronicle* by day that he was jabbing at during off hours on his site. When word got out who Banjo Jones really was, the *Chronicle* let him go.

Olafson's transgression here seems pretty clear. How could he, and by extension his paper, now be trusted by officials and readers to approach his sources fairly, once it was clear what his personal views were on the topics and people he covered? Fred Brown Jr., a past president of the Society of Professional Journalists and past chairman of national ethics, commented this way on the case: "all reporters have biases, but a good reporter never lets them show." He added: "you are not supposed to secretly attack the people you are trying to cover in an objective manner. It's not just damaging to the reporter, it's damaging to the newspaper's credibility."*

This case also points out that, contrary to what might be commonly assumed, reporters don't have free speech protections automatically from their employers, unless such rights are explicitly ceded to them.

J.D. Lasica, a senior editor for *Online Journalism Review,* thinks a looser standard for bloggers may be perfectly acceptable. "I think it's fine to pass along tips or rumors, as long as it's with the caveat of here's what I know to be true, here's what I'm not sure about and here's what I've heard from other users since my last posting," Lassica wrote in a piece that summarized the results of a panel discussion on blogging and journalism held at the University of California Graduate School of Journalism on September 17, 2002.[38] Lassica also participated on the panel, as did Gillmor. Gillmor made a pitch for retaining the traditional ethic. "I don't lose standards just because it's going online" he asserted. Another panel participant, Rebecca Blood, author of *The Weblog Handbook*, took a kind of middle ground, stating, "It is unrealistic to expect every weblogger to present an even-handed picture of the world, but it is very reasonable to expect them to be forthcoming about their sources, biases, and behavior." Her book even includes a suggested separate bloggers' code of ethics (which we briefly touched on in our discussion of ethical codes in chapter 1).

Accuracy and fairness aren't the only journalistic ethics, of course. There's also the matter of whether bloggers are also obligated to steer clear of real or perceived conflicts of interest. For instance, one blogger took some of his colleagues to task after they wrote reviews about some of the latest mobile devices introduced by Microsoft but did not inform

---

* "Web Site Targeting Local Politics Shut Down," *The Facts,* Clute, TX, July 26, 2002. Retrieved from the World Wide Web: http://thefacts.com/story.lasso?wcd=4266.

their readers that their trip to see the new gadgets was paid for in full by the company.[39]

Perhaps asking whether a blogger should specifically be constrained by traditional journalistic standards is not the right way to approach determining their exact ethical responsibilities. No matter what we call them, bloggers are *not* traditional journalists. But they are communicators with influence and as such have potentially great impact on their readers.

A basic tenet of ethics is the responsible use of power. Richard L. Johannesen, in his section on "Freedom and Responsibility," quotes John Merrill's book *The Dialectic in Journalism: Toward a Responsible Use of Press Freedom* about the tensions that inherently exist between freedom and responsibility; and that ultimately, the higher synthesis that can emerge from the tension is "the ethical use of freedom."[40]

Whether the specific standards that bloggers determine are ultimately appropriate for them, an awareness of the importance of the ethical dimension to their communication should be cultivated.

It seems clear that blogging is a phenomenon that will continue to increase and play a larger role in the entire media environment. Blogging may even morph into increasingly powerful new forms. Cell phones that take photographs and wireless personal digital assistants (PDAs) are expected to increase, and eventually these gadgets may even come equipped with a built-in video camera. At this point, bloggers will not only be able to tell readers their news but to show them the news as it happens—and even get immediate feedback from the scene. A new term has even been created to describe this kind of on-the-go blogging: moblogging, for "mobile blogging." A Web site that relies on mobloggers for its reports is called HipTop Nation (http://hiptop.bedope .com/).[41]

This raises a very interesting scenario where the readers'/viewers' feedback and commentary occur *while an event is occurring*, transforming the audience into more than just an observer. This will certainly raise a whole host of new ethical questions and concerns.

A New Model for Online Journalism?

Mark Deuze, a communications professor at the University of Amsterdam who specializes in digital media, is in general agreement with Kovach and Rosenstiel's approach (he is a member of the CCJ). However, he also finds value in the participatory potential of open journalism and suggests that the central purpose of journalism articulated by the CCJ could be expanded for the Web. "For online journalism," Deuze writes, "one could add the purpose of *offering users platforms and tools to exchange views and information needed to realize freedom and self-government.*"[42] By offering readers more interactivity and community-building features—the "platforms" and "tools"—Deuze believes online news services can help liberate and empower citizens culturally and politically, which, in turn, can invigorate democracy.

Before arriving at this conclusion, Deuze conducted extensive research on news and information Web sites. He then constructed a model that identifies four distinct types of online journalism. Deuze classifies sites according to their focus on either original editorial content or public connectivity—connecting users to a variety of external content—on the one hand, and moderated or unmoderated audience communication on the other (see Figure 3.9).[43]

Here is a summary of each classification:

1. *Mainstream News Sites:* These sites typically adhere to the norms of traditional journalism with regard to newsgathering, editing and audience relationships. As Deuze explains, they offer a "selection of editorial content and a minimal, generally filtered or moderated form of participatory communication." Many of these sites are affiliated with commercial print or broadcast media—though there are some stand-alone sites such as Salon.com and Alternet, a noncommercial, alternative news syndicator, that loosely fit this type. Content in this

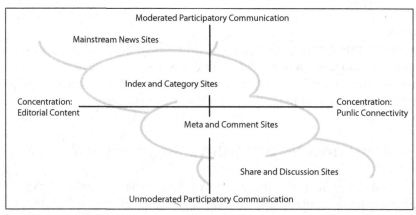

Figure 3.9 Deuze's online journalism model.

environment is a mix of original, Web-only items and material "shoveled" to the Web from the sites' offline counterparts. Web editions of newspapers such as the *New York Times, Wall Street Journal,* and *Washington Post,* as well as broadcast affiliated sites such as CNN.com and MSNBC.com, are popular examples of this type.

2. *Index and Category Sites:* Rather than publishing original news articles, the second type of online journalism is devoted primarily to linking readers to content on existing news sites. Popular examples of index and category sites are Yahoo! News, AOL News, Moreover and News Is Free. Often called aggregators, some of these sites are operated by Web portals or search engines and usually employ editors who gather, organize and annotate hyperlinks to news items about a wide range of subjects. Editors also monitor breaking news situations, maintain sections with links to the latest updates, and monitor chat sections and discussion forums. While they may not write and report regularly, many news aggregators help readers find critical information and facilitate public dialogue about important issues. Some individual Weblogs may also fall in this category.

3. *Meta and Comment Sites:* The next type of online journalism, according to Deuze's study, produces what could be called "journalism about journalism." Most sites in this category publish news, commentary and research about issues related to journalism and media production in general (Media Channel, Freedom Forum, Fairness and Accuracy in Reporting, Editor & Publisher, the Poynter Institute and the *Online Journalism Review* are prominent examples). Others, such as the Alternative Press Center and the Independent Press Association, provide extensive links to nonmainstream news outlets. Meta and comment sites most often serve as platforms for media watchdogs and critics who hope to, in the words of Deuze, "reinvigorate journalism's function of carrying on and amplifying the conversation of the people." Weblogs operated by media critics are consistent with this group.

4. *Share and Discussion Sites:* The last type of journalism consists of sites that focus on public connectivity and participatory communication with a minimum of editing and moderating. Their design most closely follows the end-to-end architecture of the Internet. As such, operators of these sites see themselves as facilitators of public dialogue, which is often centered on a specific theme such as computer technology, politics or community activism. Site users provide most

of the content by posting their own stories and ideas or commenting on the posts of other users. Examples include Slashdot, Kuro5hin, Indymedia network and group Weblogs such as MetaFilter.

With this inclusive model, Deuze recognizes the values of both traditional and nontraditional approaches to newsgathering and dissemination. He suggests that as long as a site consistently strives to fulfill the greater purpose of informing people about important issues and empowering them to engage in public debate, its content is worthy of being called journalism.

Need for More Discussion and Critical Engagement

Despite the efforts of Kovach and Rosenstiel to construct and communicate a more accessible theory of journalism based in part on traditional standards, and the efforts of Deuze to recognize the important work being done online by nontraditional news providers, there is not likely to be consensus for the answer to our question, Who counts as a journalist? Many traditionalists will continue to argue that only those working for well-edited, mainstream publications, broadcasters and Web sites are capable of producing journalism while all other outlets, valuable as they may be, produce something else entirely. From the other direction, many nontraditionalists will insist that Weblogs and sites such as Slashdot, Kuro5hin and Indymedia represent a new kind of journalism, which doesn't necessarily compete with or replace traditional journalism; rather it serves as a complement—or from the more radical perspective, an "antidote" to mainstream media power—filling in niches that traditional media don't reach. In short, the cultural divide is not likely to dissolve any time soon, though it might be bridged somewhat through thoughtful discussion and debate, not only *within* media organizations but also *between* them as well as with Web users.

Along this track, several journalists, media producers and scholars advise users to take advantage of the wide range of news offerings on the Web, from mainstream to nontraditional sites, and to hold those sites accountable for what they publish, no matter which techniques they use. Mindy McAdams, a University of Florida journalism professor, suggests that a healthy news diet should consist of a variety of carefully read sources. She specifically recommends the European custom of reading diverse political journals, which are historically more open about their editorial biases than major U.S. media, in order to bring the truth gradually into focus. Foster of the open journalism site Kuro5hin offers similar advice for Web users: "One voice is never authoritative . . . The only way to get at a clear picture of the world is to consult a variety of sources, and confront them all with your brain engaged."[44]

Summary

Since anyone with a computer and a modem can easily gather and disseminate news and opinion, the age of the Internet has raised a difficult question: Who should qualify to be called a journalist? Some define a journalist by his or her ethics and standards or by journalism's "core values." Others take a more traditional viewpoint and look at the actual function and activities of a traditional journalist, such as "filtering" news from noise, serving as an arbiter to what is important, and helping make sense of the day's events.

Another perspective is called "open journalism," a point of view that sees the Internet's capacity to provide for two-way communication as a way to expand and open up who may be called a journalist. Critics of open journalism worry about issues surrounding looser editorial standards, especially since Internet journalism relies more on input from readers; but proponents of open journalism point to this interactivity as a strength.

The emergence of the Internet as a new medium worried many of the traditionalists about the direction of the news industry. These concerns led to the creation of the CCJ, which issued a statement that purported to bind journalists together on the basis of their belief in fundamental principles of journalism.

While there are many ways to analyze the question of who is a journalist, there are practical matters that make the answer more than an academic exercise. For example, those who are designated as journalists obtain special permissions to attend news events; even more important is the fact that some states give journalists legal protections against being forced to reveal the identity of confidential sources.

A completely different publishing format that occurs on the Internet is what is known as Weblogs (or blogs): personal journals that usually consist of short news updates, links and opinions of a single individual. Weblogs became particularly prominent after the terrorist attacks of September 11, 2001, as a source of compelling eyewitness accounts and personal narratives.

Most bloggers don't consider themselves to be traditional journalists, and many even eschew the ways that modern journalism has been practiced. Some blogs are relied upon not just by casual Web users but also by traditional journalists who sometimes find leads and first-person reports by reading popular Weblogs. It's unclear whether journalism's traditional standards and ethics should apply precisely to Weblogs; in fact, a separate code of ethics has been suggested specifically for bloggers. Another ethical issue related to Weblogs surfaces when a working journalist decides to begin his own personal Weblog on the side, as what he reveals there may create conflicts in how effectively he can carry out his mainstream reporting duties.

Ultimately, there is likely to be little final or "official" consensus on the question of who counts as a journalist, though the discussion itself can be useful to surface larger issues and spur thinking about the purpose and role of all types of purposeful forms of mass communication.

Critical Thinking Questions

- Is it important to try to come up with a definition of who is a journalist? Why or why not?
- Katz says that "anybody who sees him or herself as a journalist becomes one, which is the way it ought to be." Do you agree or disagree, and why?
  - If you agree with Katz, what would be the implications regarding applying the protections of the First Amendment and its corollaries, such as the shield law? Should this matter?
  - Would Katz's argument be valid for other professions or crafts? Consider applying his method to lawyers, schoolteachers and sculptors. In which cases might Katz's statement hold true? For those that do, what does that indirectly say about journalism?
- What important function does filtering information provide to the public? What sources, institutions or individuals do you currently rely on to serve as your filters?
- If a reporter working for a traditional media organization starts his or her own blog and gets a tip on an important new story, is it ethically permissible to use it for the blog and not for the news organization? What basic journalistic ethical principle needs to be called on to make this determination?

- Which of the following factors should give one the right to call oneself a journalist? An advanced degree in journalism or communications; longtime experience in reporting/writing and/or editing; having a curious nature and a desire to communicate; having developed skills in research, writing and information verification; being hired by a traditional media organization to work as a reporter, writer or editor; starting a Weblog; contributing to someone else's Weblog; or are no factors necessary—if someone says they are a journalist, that is enough.
- What is your own definition of a journalist?

Key Terms

Blogger
Moblogging
Open journalism
Shield laws
Traditional journalism

Recommended Resources

Books

Blood, Rebecca. *The Weblog Handbook: Practical Advice on Creating and Maintaining Your Blog* (Cambridge, MA: Perseus, 2002).

Publications

Online Journalism Review. Web site: www.ojr.org.
Microcontent News. Web site: www.microcontentnews.com.

PART II
Societywide Ethical Dilemmas for Online Media Professionals

In the second part of this text, we examine in detail three areas where the Internet has brought about new and thorny concerns with society-wide ethical implications. These broad matters are in the areas of privacy (chapter 4), speech (chapter 5), and intellectual property/copyright (chapter 6).

For each chapter, we begin by tracing the development of the issue and its evolution over time, then we identify the larger ethical components and dilemmas. Finally, we discuss in detail ethical concerns specific to online media professionals and surface specific case studies to illustrate how these dilemmas have been confronted and addressed by actual online media organizations.

CHAPTER 4

Privacy

Chapter Goals:

- Review the historical roots of the concept of privacy.
- Examine international legal protections for the right to privacy.
- Review the impact of technology on privacy and the resulting legal changes to reflect technology's impact.
- Discuss specific ethical concerns regarding a user's privacy on the Internet, including the collecting and sale of personal data, surreptitious monitoring of Web usage, and judging individuals based on their Internet use.
- Explore ethical dilemmas and cases of specific concern to the online media: using discussion forums as source material, creating privacy policies for an Internet site, and posting public records on the Web.

The Concept of Privacy

Throughout history, the concept of privacy has come to mean different things to different people depending on physical context, cultural environment and personal preferences. For example, according to author and attorney Jeffrey Rosen, being watched by a neighbor without one's consent was deemed so threatening to privacy that early Jewish law had provisions to protect individuals from the "unwanted gaze" of fellow citizens.[1] Similarly, medieval English law had provisions to protect people from peeping toms and eavesdroppers.[2]

By the 18th century, the "king's men" were seen as a major threat to privacy in England and in the American colonies. Members of Parliament a's well as colonial rebels argued that soldiers of the crown should be prohibited from entering a private house and seizing personal papers.[3] Rapid technological change in the 20th century combined with social protest and Cold War politics raised new fears about intrusive government. People often feared, with good reason, that their private communications were being intercepted and monitored by state security and law enforcement agencies.

Today, participants in the "information economy" may be just as likely to be concerned about commercial data collectors as they are about government agents or curious neighbors.

The point in this short survey is to highlight the fact that privacy affects several spheres of social life, thus it is often considered to be one of the most fundamental human rights. At the same time, it is a concept without a single definition. Given this situation, scholars and advocates tend to treat privacy more like an overarching principle made up of several distinct but related concepts. These have typically been divided into four areas:

- *Bodily privacy:* The right to be protected from invasive searches of an individual's physical self.
- *Information privacy:* The right to set limits on the collection and handling of personal data such as medical records, credit reports, tax records or consumer purchases.
- *Communication privacy:* The right to communicate through terrestrial mail, e-mail, telephone or other device without being monitored.
- *Territorial privacy:* The right to set boundaries around a physical space or structure, such as a home or office, in which personal affairs can be conducted.

Our concern is primarily with information privacy and communication privacy as they relate to the Internet and the World Wide Web, though territorial privacy will come into play when we consider the monitoring of e-mail and Web use in the workplace. Before moving on to current ethical dilemmas, it might be helpful to sketch out the evolution of privacy as a legal right against the backdrop of rapid technological change.

Privacy, Technology and U.S. Law

The right to privacy is enshrined in the Universal Declaration of Human Rights,[4] which was ratified by the United Nations General Assembly in 1948, as well as several international treaties and the comprehensive privacy laws of several countries such as Canada and Australia (see Figure 4.1). However, the United States does not have a general privacy law, nor does the Constitution spell out any explicit right to privacy, though the Fourth and Fifth Amendments deal with some aspects of territorial privacy—protection against unreasonable search and seizure—and a 1965 Supreme Court decision raised the status of the right to privacy in marital relationships (bodily privacy) to the level of constitutional protection (*Griswold v. Connecticut*), which was expanded in *Roe v. Wade* (1973). As such, information and communications privacy have been protected by "sectoral laws", which govern specific areas, such as financial or medical records, and are often passed in response to public outcry over government or commercial practices or the introduction of new technology. Advocates

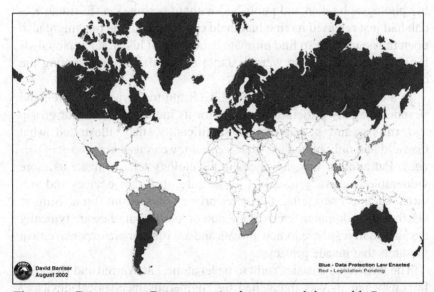

David Banisar
August 2002

Blue - Data Protection Law Enacted
Red - Legislation Pending

Figure 4.1 Privacy map: Data protection laws around the world. *Gray countries:* Comprehensive data privacy laws pending; *Black countries:* Comprehensive data privacy laws enacted; *White countries:* No comprehensive data privacy laws.

who support comprehensive national privacy laws argue that the sectoral model is often ineffective because protections tend to lag far behind technology.[5]

The Telegraph and the Camera

Looking back, it appears that privacy concerns and technology first converged with the emergence of the telegraph as a major communications medium. In the 1860s, James Garfield, a congressman who later became president, conducted hearings into the practices of Western Union, the sole telegraph operator in the country. Robert Ellis Smith, an attorney and privacy historian, reports that there were "no guidelines on protecting the confidentiality of the sensitive information" being handled by Western Union.[6] Garfield's efforts to warn about the threats posed by new technology and corporate monopolies did not lead to the establishment of any new privacy laws, however.

Two decades later, Samuel Warren and Louis Brandeis, two Boston attorneys, became alarmed at the possibility that one might have his or her photograph taken and published without permission (Eastman Kodak had just released its first handheld camera), and that one might also open the newspaper to find intimate details about his or her personal affairs (rumor and gossip were a staple of the daily press) among the day's news.

In a seminal legal essay called "The Right to Privacy,"[7] Warren and Brandeis harshly criticized the press for its increasing thirst for gossip and rumor; and perhaps more significantly, they illustrated what Garfield had only hinted at—that *technology changes the nature of privacy*. Put another way, advances in technology tend to make us more vulnerable to privacy invasion. To be sure, there are devices and services, such as encryption, that offer privacy protection. But as Simson Garfinkel, a computer scientist and author, points out, they are typically developed in response to new threats and are often more expensive than products that invade privacy.[8]

The individual's basic "right to be let alone," as Warren and Brandeis famously put it, was threatened by "numerous mechanical devices," such as the camera, which could be used to expose intimate moments. Unless the legal protections kept up with technology and provided some remedy for "the evil invasion of privacy," they argued, one's personal conversations and likeness could not be protected. On a deeper level,

Brandeis, who later became a Supreme Court justice, also argued that privacy was so essential to human dignity and autonomy that it should be protected at the constitutional level in the same manner as freedom of expression and freedom of association.

Warren and Brandeis' 1890 warning had little immediate impact at the national level. However, they had also urged the states to step in where the federal government would not, and in the decades that followed, many states began to recognize a citizen's right to recover damages for the use of personal information or images without consent.

Public vs. Private Citizens

The two attorneys also helped spur debate over which issues were of "public interest" and whether or not some citizens—officeholders and government administrators, for example—were entitled to the same protections from the press as other citizens. Warren and Brandeis allowed that elected officials had surrendered a measure of privacy by entering politics but only with respect to "peculiarities of manner and person" or other information connected to their fitness for office—they still had a right to protect intimate details of events and relationships not related to public business.[9]

As for private citizens, Warren and Brandeis believed they deserved a higher degree of protection but not absolute protection. "The right to privacy does not prohibit any publication of matter which is of public or general interest," they wrote.[10] For example, the inventor of a life-saving medical procedure could expect information about his experiments and research habits to be exposed. Conversely, a convicted criminal could expect private information related to his or her background to be published.

Despite these allowances, journalists as well as many jurists have traditionally opposed the idea that laws—be they state or federal statutes— or the courts should decide which persons or issues are in the public or general interest. Newsgathering and reporting, they've argued, are fully protected by the First Amendment, thus decisions about what facts are suitable for publication should be made in the editorial offices, not in the courtroom or legislative chamber.

In 1940, the case of William Sidis, a former child prodigy who had become a reclusive office clerk, put Warren and Brandeis' theory to the test.[11] At issue was whether or not Sidis' privacy was invaded by the publication of a 1937 "where are they now" article in *The New Yorker*. The appeals court judges hearing the case admitted that the article was a "ruthless exposuré of a once public character" who was determined to fade into obscurity. By Warren and Brandeis' standards, they agreed, Sidis' privacy had been invaded. However, they believed those standards to be too strict. Sidis' intellectual achievements as a youth, they argued, had garnered much public attention, which might cause people to be curious about his current activities—to want to know whether or not "he had fulfilled his early promise"—and as such, the popular interest trumped Sidis' efforts to protect his privacy.

It is important to draw a clear distinction between privacy and libel or slander. Privacy torts are concerned mostly with intrusion into the private sphere to gather news or the unauthorized dissemination of one's likeness as well as the reporting of truthful, but embarrassing, personal information. Libel law, on the other hand, is concerned with the reporting of false information about a person, which might damage his or her reputation. As with privacy, though, libel law is also concerned with the question of whether one is a private citizen or public official. For example, in the landmark *New York Times v. Sullivan* (1964) decision, the Supreme Court ruled that the First Amendment protected criticism of public officials containing false information, so long as information was not presented with actual malice—that is, reckless disregard for the truth. In *Gertz v. Welch* (1974) justices ruled that private citizens need only prove negligence on the part of the publisher to win a libel case.

The Telephone

By the early 20th century, the telephone had emerged as a popular and accessible communications medium. It proved to be an immensely popular technology but was also a medium that could be easily tapped so that a third party could eavesdrop without being detected.

In 1928, the Supreme Court addressed the question of whether or not the government should be allowed to gather evidence by secretly tapping the telephone of a crime suspect. In *Olmstead v. the U.S.*, the court ruled 5–4 that secret wiretapping did not constitute an unreasonable search and seizure because agents never physically entered a private office or dwelling, nor did they confiscate any personal papers or other effects. Brandeis, who by this time had become a justice on the high court, argued forcefully in a dissenting opinion that the majority had erred by interpreting the Fourth Amendment too narrowly. Under a broader interpretation, one that took modern technology into account, Brandeis asserted that wiretapping could clearly be seen as an invasion of privacy. He argued further that secretly tapping the phone of the suspect amounted to "the tapping of the telephone of every other person whom he may call, or who may call him." Given all this, Brandeis concluded, wiretapping was a far stronger "instrument of tyranny" than an unjustifiable physical intrusion.[12]

Nearly four decades later, in 1967, amid growing complaints about increasing government surveillance, the Supreme Court reversed the *Olmstead* decision and asserted that "the Fourth Amendment protects people not places,"[13] thus individuals had the right to conduct private telephone conversations without being monitored. Congress responded to the court's ruling one year later by passing a law that authorized wiretapping only under certain conditions and with a judge's order. In 1986, the Electronic Communications Privacy Act (ECPA) extended the wiretapping provisions to protect e-mail messages from unauthorized surveillance.

Computer Databases

During the 1960s, the development of powerful mainframe computers with remote terminals allowed businesses as well as government agencies to efficiently gather, store and retrieve huge amounts of personal information. The new information architecture plus the sheer volume of data being generated raised new concerns about data accuracy, security and the practice of sharing databases by federal agencies—or the selling of databases by businesses—for the purpose of compiling dossiers on citizens.

On the commercial side, the Fair Credit Reporting Act (FCRA) was passed in 1970 in response to numerous cases in which data used to determine an individual's credit rating—such as bankruptcy filings, employment history and bill payment records—were discovered to be inaccurate, yet still being sold by credit bureaus to other businesses. The act limited the distribution of credit reports and gave citizens the right to access their records and challenge inaccurate data. In subsequent years, the FCRA has been updated to expand consumer rights, though not as much as many privacy advocates desire.[14]

Government record keeping also generated public criticism when in the early 1970s it was revealed that the FBI was not obliged to guarantee the accuracy of information it stored and shared with other law enforcement agencies. Congress responded by passing the Privacy Act of 1974. The measure limited data gathering and sharing by federal agencies and required them to apply "fair information practices" (FIPs), which consist of the following principles:

1. There must be no personal data or record-keeping systems whose existence is secret.
2. A person must be able to find out what information is contained in his or her record and how it is used.
3. A person must be able to prevent information obtained for one purpose from being used or made available for other purposes without his or her consent.
4. A person must be able to correct or amend a record of identifiable information.
5. Any organization creating, maintaining, using, or disseminating records of identifiable personal data must assure the reliability of the data for its intended use and must take precautions to prevent misuses of the data.[15]

Despite the strong provisions, many critics say the absence of a federal privacy oversight agency has made it difficult to enforce the act.[16] The act also does not prevent agencies from "outsourcing" data gathering and storage to private firms not covered by the law. Government agencies are now increasingly buying huge volumes of personal data from commercial data collection companies—and in some cases selling data to private businesses. Marc Rotenberg, director of the Electronic Privacy Information Center, told the *Wall Street Journal* that agencies are using these partnerships to make an "end run around the Privacy Act."[17]

One company in particular, ChoicePoint Inc., has multimillion-dollar contracts to gather, sort and package information for the Internal Revenue Service (IRS), FBI, Immigration and Naturalization Service (INS), U.S. Marshals Service, Health Care Financing Administration and several agencies at the state level. As with earlier databases, the accuracy of ChoicePoint's information has come into question. In January 2001, the NAACP sued the company on the grounds that it supplied erroneous data to the state of Florida—data that was used to purge names from the state's voter rolls before the 2000 presidential election. The company admitted that some of its data was inaccurate, but it also claimed that it warned state officials of the need to verify the information.[18]

The Internet and the Web

While some government agencies were building massive databases in the 1960s and '70s, other agencies, the Department of Defense in particular, were building a revolutionary new communications system, which eventually became known as the Internet. This system could best be described as a decentralized network of computers using a set of standardized protocols to break data into packets so that it could be sent efficiently from one remote computer to another.[19]

E-mail, which was developed in 1971, was one of the first applications of this new system. For nearly two decades it was a communications tool used mainly by government and academic researchers. That all changed in the early 1990s when a new application, the World Wide Web, was introduced. Its creator, Tim Berners-Lee, refers to it as a communications platform that "rides on top" of the basic Internet communications infrastructure.[20] Unlike e-mail and other early applications, which did not create a permanent space for information to exist, the Web enabled users to store data files on computers called "servers." Users at remote terminals could then download this data or link to it using hypertext.

Some of the data now available on the Web comes straight from records in government databases that have been released into the public domain—arrest records, motor vehicle registrations, tax liens, and so forth. In theory, these documents have always been available to citizens. However, before the Web, they resided in "practical obscurity." That is, residents had to travel to courthouses or agency offices, fill out an application and pay fees to obtain printed copies of the records. Now, from the comfort of one's home, office or public library, access is only a few mouse clicks away. Journalists, researchers and citizen-activists have benefited greatly from the new accessibility. They say that the ability to retrieve public records in a timely manner helps them inform the public, thus making government more directly accountable to citizens.

Several state and county governments have published documents and public records on the Web. In Hamilton County, Ohio, for example, most of the county's court records can be easily searched and down-loaded from a county-run Web site. Employers have used the site to per-form background checks on job applicants, while some criminals have used it to retrieve Social Security numbers. Some residents have even looked up their neighbors' divorce proceedings. Along with attracting hundreds of thousands of users, the site has also generated numerous complaints from those whose records have been exposed.[21]

Privacy advocates argue that some of the data found in these online records, such as addresses, place of employment and Social Security numbers, can allow anyone with a gripe to steal an identity or easily track and harass a fellow citizen. In 1999, a New Hampshire woman was shot and killed by a stalker who had located her with the help of an "on-line detective" who found the woman's Social Security number and workplace address in a database of public records.[22]

Some states have taken action or are considering proposals that would prevent certain types of personal information from getting on the Web. In California, for example, the governor put a freeze on the sale of the state's birth records, which had been posted on privately operated genealogy Web sites.[23] An Ohio lawmaker has also introduced a bill that would keep certain sensitive personal information in the public domain from being posted online.[24]

Journalists and lawmakers concerned about balancing privacy with the public's right to know worry that these measures would circumvent open records laws by creating a two-tiered system of access—one for offline information and another for online records. Given the likelihood that most public records will find their way online, they argue, it's the concept of the public domain that needs to be re-evaluated. The court clerk in charge of the Hamilton County, Ohio, database has proposed that certain records such as "juvenile court transcripts, financial state-ments, and psychiatric reports should be exempt from open records re-quirements offline as well as online."[25] Under his model, all records classified as public would be made easily accessible in both traditional and digital formats.

Government Surveillance

The Internet and the Web not only facilitate easy data storage and re-trieval; they also allow for instantaneous communication among widely dispersed individuals. As with telephone conversations, those messages can be monitored by law enforcement and other government agencies. This is largely due to a 1994 law, which mandates that digital telecommunications carriers, who control the pipelines to the Internet, design their systems to be tappable by law enforcement. In other words, the architecture of the system must be compatible with surveillance software and other monitoring programs.

Under a 1968 wiretapping law and the ECPA, police were required to obtain a court order and meet certain legal conditions before monitoring or intercepting Internet communications. However, those provisions were weakened by the USA PATRIOT Act, which was passed in the wake of terrorist attacks in the United States on September 11, 2001. Law enforcement can now demand that ISPs install a surveillance program called Carnivore, which harvests e-mail and other online communications, based on a prosecutor's certification that information it would collect is relevant to an ongoing terrorist investigation. Courts would also be able to issue "roving wiretaps," which would give law enforcement agents the ability to monitor a variety of communications devices that a suspect might use—cell phones, PDAs or computers in a public library, for example.[26]

On this last point, a survey conducted in 2002 by a library science professor indicated that 85 libraries in the U.S. were approached by government agents in the months following the terrorist attacks and asked to turn over certain patron records. The same survey reported that over 100 libraries have instituted more restrictive policies for Internet use while 77 indicated they had monitored patron activities since September 11th.[27]

Many privacy advocates have come out strongly against the USA PATRIOT Act, arguing that it undermines traditional privacy protections by weakening, and in some cases removing, judicial oversight of surveillance procedures, which, they contend, is vital to protecting citizens from unreasonable search and seizure. Supporters of the act claim the new surveillance powers are needed to help locate domestic terrorist cells that are adept at using digital communications technologies to coordinate movements and strategies.

Consumer Surveillance and Profiling

Along with the attention focused on the USA PATRIOT Act, many are also concerned about Internet monitoring conducted by commercial interests. After the basic Internet infrastructure was privatized in the mid 1990s, numerous businesses went online in search of customers who were simultaneously logging on in search of news, consumer information and other services. As Internet users have resisted paying directly for access to content or services, many of those early Web businesses collapsed. Others have remained online, experimenting with a variety of revenue models in hopes of becoming profitable.

The most dominant model of late has two variations: (1) the collection of personal information in exchange for direct access to information and services; and (2) the gathering and aggregating of customer information into "consumer profiles," which help advertisers predict a Web user's interests and purchasing habits so that they can deliver specially targeted marketing messages. Put another way, personal data has become the primary "currency" in the Internet economy.[28]

Personal data is often gathered openly when users fill out registration forms to enter a Web site, register for an online contest, or set up an e-mail account. Many sites and ISPs sell membership information to online marketers and other businesses that use the data to construct profiles. However, not all of them disclose this practice. Users who liberally divulge personal information such as their age, income and educational background are likely to become the targets of online marketing campaigns often consisting of unsolicited e-mail messages called "spam." In 2001, the European Union passed a law that prohibits marketing by e-mail without user consent. As of late 2002, the United States has not adopted a similar law.[29]

Customer information is also collected when users "agree" to let a Web site install small text files called "cookies" on their computers. These files enable the site to "recognize" users each time they visit and to keep track of what they do while surfing various pages, like capturing terms typed in a search engine. Web sites, in effect, use cookies to build a profile of each user based on his or her online behavior. This can make Web surfing and shopping quick and convenient—for example, when cookies are used to hold shopping cart information until one is ready to make an online purchase. However, the use of cookies gets a bit more complicated when the information gathered is sold to third parties without a user's consent, or when Web sites allow third-party cookies— typically those belonging to advertising networks—to be surreptitiously installed on a user's computer.

Most Web browsers do not automatically alert users when third-party cookies are about to be installed unless the user changes the default

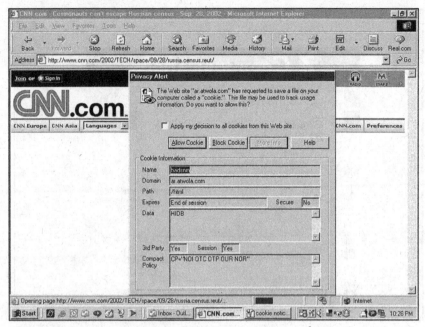

Figure 4.2 Third-party cookie alert in Microsoft Internet Explorer.

preferences (see Figure 4.2). When the alert option is selected in the Microsoft Internet Explorer browser, for example, the user is given the name of the Web site trying to install the cookie and a warning that the cookie may be used to track Web site usage. It may also allow the user to block the cookie. However, many Web sites do not allow users who block cookies to enter any portion of the site. For this reason, many users do not use the alert feature.

In more recent versions of Internet Explorer, users who don't want to bother with alert boxes every time a third-party cookie is about to be installed have the option of choosing a privacy protection option (low, medium, medium-high, high, or "block all cookies") that suits their needs (see Figure 4.3). The browser will then monitor Web sites and either accept or reject cookies according to the level chosen. Medium privacy, for example, blocks third-party cookies that use personally identifiable information without a user's "implicit consent."

Information is also gathered covertly by Web sites through the placement of "Web bugs" on a user's computer. A Web bug is a nearly invisible graphic on a Web page or in an e-mail message that is typically programmed to gather the computer's IP (Internet protocol) address, the

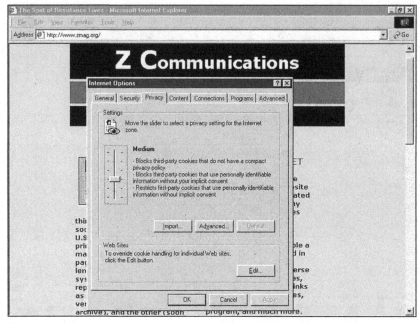

Figure 4.3 Privacy options setting in Microsoft Internet Explorer.

type of browser being used, and the time they opened the e-mail or connected to the Web page. As with data gathered by third-party cookies, this information is most often used by marketing companies to build consumer profiles.

Consumers and privacy advocates upset about deceptive profiling practices have filed numerous lawsuits against Web sites, ISPs and advertising companies. In addition the Federal Trade Commission (FTC), which has jurisdiction over online commerce, has launched several investigations. In response, some companies have changed their profiling practices. In late 2001, for example, DoubleClick, a major advertising company, closed its online profiling division.[30]

The FTC also released a survey in 2000, which indicated that online services and ad agencies were not doing enough to protect consumer privacy. The agency concluded that self-regulation had achieved only "limited success" and recommended that comprehensive privacy legislation based on FIPs be adopted.[31] However, in 2002, the agency backed away from that recommendation and endorsed a new survey conducted by the Progress and Freedom Foundation, which suggested that self-regulation was working and that new laws were not needed.[32]

In keeping with the sectoral approach, Congress has not yet adopted a comprehensive federal law protecting online privacy (as of late 2002, several bills offering broad protection had been introduced), opting instead to update existing privacy laws to protect specific types of data, such as medical records, online. There has been one small exception to this approach, the Children's Online Privacy Protection Act (COPPA), which went into effect in 2000 and prohibits the collection of any information from children under the age of 13 without parental consent.

Though federal lawmakers have not enacted sweeping privacy protections, their counterparts at the state level have begun to make moves in that direction. In Minnesota, a new law will put strict limits on the use of personal information collected by ISPs and regulate the transmission of spam.[33] New online privacy measures have also been introduced in several other states, including Hawaii, Arizona, New Jersey, Missouri and Massachusetts.[34] Given the Internet's global reach, state laws are likely to raise difficult jurisdictional and enforcement questions.

Many privacy advocates applaud the action by the states, while others continue to lobby for comprehensive federal regulations more in line with European,[35] Canadian,[36] and Australian[37] data privacy laws. Simson Garfinkel, author of the book *Database Nation*, sums his argument up this way: "Without government protection for the privacy rights of individuals, it is simply too easy and too profitable for business to act in a manner that's counter to our interests."[38]

At the other end of the debate, several marketing associations and many online businesses believe that sweeping, government-imposed regulations are not the answer. Services that don't respect the privacy demands of their customers, they argue, will lose business. From this point of view, the marketplace rewards those who protect customer privacy while punishing offenders.

Online Privacy: Ethical Dilemmas

Collecting and Selling Personal Data

Given the economics of the Internet economy—personal information being the "currency" of the online economy—and a political climate that favors industry self-regulation, it may be profitable and perfectly legal for online businesses, including online news organizations, to package and sell customer information without the customer's knowledge or permission. Furthermore, digital technology allows businesses to use "invisible" data collection techniques (e.g., cookies and Web bugs), which most Internet users know very little about.[39]

But are such practices ethical for e-commerce sites? How about news Web sites—is it wrong for companies doing journalism to get into the business of secretly gathering and selling customer information? Should they even track the online behavior of visitors for internal marketing purposes?

After surveying public attitudes about online information gathering, Lee Rainie, director of the Pew Internet and American Life Project, told members of Congress in 2001 that Internet users "viscerally oppose the ideas of online tracking and profiling"[40] because they see them as invasions of privacy. Users feel that secret monitoring destroys the ability to surf the Web anonymously, which they consider to be a fundamental right. At the same time, users also express a desire to participate in online commerce and would be willing to share some personal information if they could control how it would be used. Thus if Internet users could adopt a "golden rule" for online privacy it would be, "Nobody should know what I do on the Web or anything else about me unless I say so."[41]

This sentiment is very much in line with the principle of "informed consent," which underpins FIPs that are applied to government agencies in the United States. In the absence of a legal mandate, numerous privacy advocates and consumer groups argue that commercial Web sites and other online businesses should, at the very least, institute privacy policies that adhere to the following FIPs recognized by the Federal Trade Commission (FTC):

1. *Notice:* The consumer must be notified, in clear language, that information is going to be collected and be informed as to the purpose for data collection and the conditions under which the information may be shared or sold.
2. *Choice:* The consumer must be able to decide whether or not personal information can be collected and shared.
3. *Access:* The consumer must have reasonable access to his or her information and be able to change or delete data.
4. *Security:* Data collectors should be responsible for maintaining the security of information.[42]

Others suggest that in order to provide maximum protection and satisfy consumers' wishes, commercial FIPs should also include the following two principles recognized by the Organization for Economic

Cooperation and Development (OECD) and most European nations. First, a minimization principle, which would prevent the collection of any personal data beyond that which is needed to conduct a transaction. To give an example from the offline world, grocery and department stores do not need to know your phone number in order to complete a cash or credit card transaction; therefore, clerks should not be instructed to ask for it. Similarly, online businesses should not gather any piece of information, such as a record of Web links accessed during each visit, unless it is required to complete a purchase or deliver a requested service. Second, limits on the use of the data. Personal information should not be used for any purposes inconsistent with those listed in the privacy notice. For example, if a Web site states that information gathered by cookies will only be used to monitor items placed in a shopping cart, it would be improper to use that data to build a profile for future marketing campaigns.[43]

Privacy advocates contend that these principles are designed to empower individuals by giving them information about data collection practices as well as opportunities to choose what personal information they will divulge online and what they will keep to themselves. If commercial Web sites wanted to allay the fears of customers and build trust, it is argued, they would adopt all six FIPs. But there is no record of any U.S.-based commercial online service doing so. Partly this is due to the FTC's recognition of the first four FIPs, which established the recommended protocol for privacy practices in America. However, even when online businesses accept the FTC-supported FIPs, they often disagree with privacy advocates as to the proper way to interpret and implement them. This is especially true of the second principle—consumer choice.

Consumer Choice: Opt Out or Opt In?

At issue is the question of whether "opt out" or "opt in" policies are most consistent with the intent of FIPs and the privacy expectations of Internet users. Also up for debate is the question of which policy is most ethical.

In theory, both schemes give consumers the ability to choose whether or not they will disclose personal data or allow online services to send them marketing messages. However, in practice, they are quite different. For example, opt-out policies assume that the *user approves* the collection and use of information or the sending of spam—"yes" is set as the default preference. If the user feels otherwise, he or she must take the initiative to contact the company and opt out of the program.

By contrast, opt-in policies assume that the *user does not approve* the collection or sale of personal information or the sending of marketing messages—"no" is the default preference. Even if the user feels otherwise, the data collector must take the initiative to obtain permission before collecting information. This would also apply if data were being put to a new use or being given to additional third parties. In other words, each time the Web site or online service changed its data collection practices, they would have to notify customers and ask permission before implementing them.

Privacy advocates argue that opt-in policies are consistent with the Internet user's "golden rule" and, hence, more ethical. Since the collector is attempting to get personal information in addition to observing and recording what a customer does online, often with the intent of selling the resulting data for a profit, the collector should take the initiative to give notice and obtain permission before data is gathered and/or sold or before marketing messages are sent.

Many online services and marketing agencies, on the other hand, argue that collectors and customers should share the burden more equally —that customers should take some proactive steps, such as researching data collection techniques and opting out of policies they don't like in order to protect their privacy. Direct marketing groups have even released studies claiming that opt-out schemes provide as much, if not more, privacy protection than opt-in policies. The studies, as one might expect, have been criticized by many online privacy advocates.[44]

Consumer Choice and Implied Ethical Contracts

Are opt-in policies more ethical than opt-out policies? In searching for an answer, it might be helpful to measure the two policies against the principle of the "implied ethical contract."[45] This refers to the unspoken assumptions and expectations that characterize many forms of public communication. What is the expectation that users have now about what happens to the personal data that they provide to a Web site—that it is fair game to be sold to other businesses for marketing people? That it may be used for marketing but only by that Web site? That it will only be used for the explicit purpose for which it was provided?

An even more basic implied ethical contract is that we expect that others will speak truthfully—as much as they can ascertain the truth—to us, unless we have some compelling reason to be skeptical. We come into most conversations with the assumption that truth will be spoken and that lying will be the exception. The default, in other words, is set on "trust."

If we agree that a Web site, or any other business, has an implied ethical contract with its users, we would expect the Web site to communicate truthfully to the user and be up front about all of its practices related to the collection and use of personal information. Deceptiveness on the part of the Web site would violate the contract. Sometimes Web sites that have opt-out policies will change the way they wish to use the data. Many Internet users and privacy advocates find this to be deceptive in that the Web site doesn't approach the customer beforehand to give notice of policy changes and ask permission to use data for new purposes. Could this be construed as a violation of the standards of openness and truthfulness that underpin an implied ethical contract?

A scenario very similar to the one just described has actually happened on more than one occasion, most notably in March 2002 when the popular Web portal Yahoo! suddenly switched from an opt-in to an opt-out privacy policy. A short statement tucked inside a long e-mail message about Yahoo!'s commitment to privacy protection notified customers that the company would be changing their personal settings to indicate that they wanted to receive marketing messages via e-mail, terrestrial mail and telephone. All users, even those who had previously refused to accept marketing messages, were told they had 60 days to change their preferences back to the original setting, which they could do by manually clicking "no" on 13 products listed on a marketing enrollment page.[46]

Perhaps most troubling for privacy advocates and numerous Yahoo! consumers was the fact that the new policy was considered to be consistent with the privacy guidelines of an organization called TRUSTe, which had certified Yahoo!'s privacy statements with its "trustmark" seal of approval. Only one year earlier, TRUSTe had criticized the Web auction site eBay, also one of its seal holders, for making a similar change to user preferences. Some Yahoo! customers have complained that TRUSTe is backsliding into the role of industry apologist, which betrays its stated mission to be an independent privacy watchdog.[47] The organization has, in fact, been the subject of numerous complaints from privacy advocates dating back to 1999.[48]

For their part, TRUSTe officials insist that they are holding Web sites to high standards and contend that Yahoo! wanted to implement even more substantial policy changes, only to be reined in by TRUSTe's oversight process.[49] A review of the organization's seal of approval program indicates that TRUSTe's minimum standard for consumer choice is the opt-out scheme, which Yahoo! had switched to, albeit in a manner that upset many customers.

This raises several ethical questions. For starters, did Yahoo! violate its implied ethical contract with users, especially those who had originally chosen not to participate in marketing campaigns? If so, is TRUSTe complicit in that violation by virtue of its support for the policy? And what good is a seal of approval if it's okay for a Web site carrying it to change customers' privacy preferences? On this last question, some customers and privacy experts argue that the TRUSTe seal may actually work against privacy protection[50] by creating the illusion that a Web site adheres to privacy standards that are consistent with the Internet user's golden rule, when in reality it struggles to meet the conditions of FIPs. At least one privacy expert suggests that seeing the TRUSTe logo on a Web site "may induce people to make information disclosures that they would otherwise avoid."[51]

News Sites and Privacy Policies

It's worth noting that among TRUSTe's 2000 seal of approval holders are dozens of news sites, including two of the Web's most popular, the New York Times.com and MSNBC.com. While the privacy policies of these outlets might not measure up to a strict interpretation of FIPs— at least as defined by most privacy advocates and numerous technology specialists[52]— they may be leagues ahead of the rest of the industry for the simple fact that they post a privacy notice on their Web sites.

In the summer of 2001, the *Online Journalism Review (OJR)* published the results of an online privacy study conducted by the University of Southern California and the University of California at Berkeley, which indicated that news media Web sites were doing a poor job of disclosing their data collection practices. Here are some of the study's highlights:

- 38 percent of the sites surveyed were providing some form of disclosure. Only 28 percent were posting comprehensive privacy policies.
- 65 percent of the sites surveyed were collecting personal information from users, such as e-mail addresses or credit card numbers, through vehicles like online registration forms.
- 68 percent of media sites surveyed used cookies to track users' movements. A majority of those sites were not disclosing the use of cookies. Among sites that were posting comprehensive privacy policies, 34 percent failed to reveal their use of cookies.
- 82 percent of the sites using cookies were also allowing third-party cookies to be placed on users' computers. Most of these sites were not disclosing the use of third-party cookies.[53]

In addition to these findings, the study revealed that the disclosure rate of 38 percent for news sites is sharply lower than the 88 percent disclosure rate for general commercial sites reported by the FTC. To be sure, not all of those sites are following the golden rule; however, they are at the very least posting some sort of privacy notice.

Larry Pryor and Paul Grabowicz, the study's principal investigator and co-investigator, suggested that the failure to give notice may be driven by the fear that prominent disclosure of data collection techniques will scare users away, thus depriving the site of valuable personal information, which could be converted into much-desired revenue. Web sites that take this approach, according to Pryor and Grabowicz, do so at their own peril. In an *OJR* editorial published the same day as the study's results, the two men urged online news organizations to take action to protect user privacy or risk losing credibility with the public. "If the online news industry continues to ignore this dismal performance," they wrote, "brands and credibility, built over many years by news organizations, will suffer."[54]

Given the fact that news organizations often demand—on behalf of the public—transparency and accountability from government and industry, we might ask whether or not they should adopt standards of privacy protection that are at least equal to or greater than the standards found in other online businesses, or those found in government. Wouldn't it make sense for news sites to live up to standards they expect others to abide by? And in keeping with our theme of implied ethical contracts, what could be said about a news organization that participates in the tracking and profiling of customers without giving notice? Does this violate an unspoken assumption that the user can trust the news organization to be open and truthful?

Teen Web Sites

As mentioned previously, the COPPA prohibits commercial Web sites and online services from collecting personal information from children under 13 without parental consent and proper notification. However, there is no law protecting older teens from Web sites and Internet portals that collect personal information in exchange for access.

This worries many privacy and child advocacy groups who contend that teens are particularly susceptible to being tracked and profiled without their knowledge because, on average, they use the Internet more frequently and much more immersively than their parents. That is, they surround themselves with media, often listening to audio files, gathering information for homework assignments, checking news and entertainment sites, and using chat rooms or instant messaging services—all at the same time. In addition, many teens work after school and on weekends, adding to their spending power and making them popular targets for marketers.[55]

A study of teen-oriented Web sites conducted in 2001 by the Center for Media Education (CME), a media literacy and advocacy group, uncovered a particularly innovative and startling example of the aggressive information-gathering tactics aimed at teens. Users of the now defunct Web site Thirsty.com were routinely encouraged to provide "both required and optional personal information in return for 'Thirsty points'," which could be redeemed for gifts and special bonuses. In addition to e-mail addresses, zip codes, mailing addresses and telephone numbers, teens could also earn points by submitting a photograph of themselves.[56]

Along with advocating greater parental involvement and media education programs in the schools, the CME report urged the operators of teen-oriented Web sites and services to take greater responsibility for protecting the privacy of their users. "At the very least," they concluded, "market researchers should be more truthful in notifying teens about their data collection practices and offer teens the option to delete any personal information that might have been collected."[57] This would bring the sites closer to the standards of the FIPs, but is that enough, considering the ages of users? Should the operators of teen sites be expected to follow higher standards than those of adult sites? How about news sites—should they implement special policies to protect the privacy of 13- to 17-year-olds?

Policy Language: Less Legalese?

Aside from concerns about privacy protection on teen-oriented, news and e-commerce sites, some Internet users and privacy advocates have noted that existing online privacy policies, no matter the genre, are often filled with boilerplate legalese, making them difficult to read. Is it enough for sites to give legal notice of their data gathering, or should they take extra care to word their statements in plain language that is understandable to those not accustomed to reading legal contracts, which is just about everybody?

At least one major Web site has been praised for developing an innovative way to explain its privacy policies. The online auction house eBay has added a "privacy chart" to its privacy notification pages (see Figure 4.4).

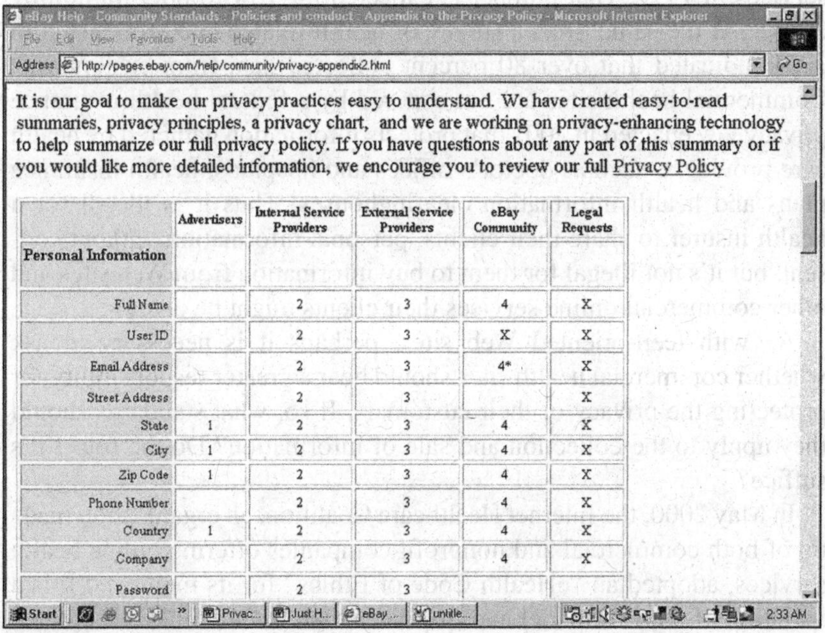
eBay Help : Community Standards : Policies and conduct : Appendix to the Privacy Policy - Microsoft Internet Explorer

Address http://pages.ebay.com/help/community/privacy-appendix2.html

It is our goal to make our privacy practices easy to understand. We have created easy-to-read summaries, privacy principles, a privacy chart, and we are working on privacy-enhancing technology to help summarize our full privacy policy. If you have questions about any part of this summary or if you would like more detailed information, we encourage you to review our full Privacy Policy

	Advertisers	Internal Service Providers	External Service Providers	eBay Community	Legal Requests
Personal Information					
Full Name		2	3	4	X
User ID		2	3	X	X
Email Address		2	3	4*	X
Street Address		2	3		X
State	1	2	3	4	X
City	1	2	3	4	X
Zip Code	1	2	3	4	X
Phone Number		2	3	4	X
Country	1	2	3	4	X
Company		2	3	4	X
Password		2	3		

Figure 4.4 eBay privacy chart.

According to an introductory paragraph, the chart "describes the ways in which we use your information and allow others to use your information."[58] More specifically, it lists the various types of information eBay collects (personal, financial, shipping, transaction and client related—that is, gathered by cookies) and indicates whether that data is shared with advertisers, internal service providers, external services providers, other eBay users, or law enforcement.

Making Judgments Based on Internet Behavior

Aside from concerns about commercial tracking and profiling, Internet users are, perhaps, more concerned that personal information may be used for other, more harmful purposes. For example, Rainie's surveys indicate that Web users are afraid that insurance companies will learn about their health concerns based on data gathered by search engines and medical-related Web sites and, as a result, change or cancel their policies.[59]

This is a genuine concern as an estimated 34.7 million adults in the United States went online in search of health information and medical services in 1999. That number is expected to grow to 88.5 million by 2005.[60] At the same time, a survey of health-related sites conducted in 2001 indicated that over 80 percent collected data from customers.[61] Commercial health sites are not covered by a federal health insurance privacy law enacted in 2001 that protects information gathered by health care providers, such as doctor's offices and hospitals, health insurance plans and health information clearinghouses. Thus it is illegal for a health insurer to share their clients' personal information without consent, but it's not illegal for them to buy information from Web sites and other commercial online services their clients might be using.

As with teen-oriented Web sites, perhaps it is necessary to ask whether commercial health sites should bear a greater responsibility for protecting the privacy of their customers. If so, what standards should they apply to the collection and sale of information? Do the four FIPs suffice?

In May 2000, the Internet Healthcare Coalition, an organization made up of both commercial and nonprofit companies offering online health services, adopted an "eHealth Code of Ethics" for its members. It is a variation of the FIP model and mandates that "sites should not collect, use, or share personal data without the user's *specific affirmative consent*."[62] This is consistent with opt-in data collection policies.

Inaccurate Data

Many Web users also fear that others, such as credit services or even prospective employers, will make judgments about them based on inaccurate data in an online profile. This, too, is a very real concern, according to Larry Ponemon, a well-known industry privacy auditor who claims that there is an 85 percent error rate in customer profiles.[63] This astonishing number underscores the danger in putting too much faith in digital tracking technology and large databases.

Journalists who use online databases, both commercial and governmental, to conduct research ought to exercise caution. Ponemon and others have noted that data transfers, computer malfunctions and operator mistakes can lead to significant errors and make data mining a risky adventure. Reporters and researchers are advised to use sites with a reputation for accuracy and to always check other sites to corroborate data. Old-fashioned methods such as telephoning sources and making personal visits to courthouses or government agencies are also recommended in cases where data appears to be suspect.[64] Chapter 8 examines ethical issues related to online reporting.

Even with accurate data in hand, some privacy advocates caution against judging others based solely on their Internet behavior. Jeffery Rosen, an associate professor of law, George Washington University, and author of the book *The Unwanted Gaze*, asserts that information about a person is too easily confused with "true knowledge," which is formed through "a slow process of mutual revelation" and should be the basis for making judgments. When personal information, such as a list of books we've read or Web sites we visited, is passed on to those with no intimate knowledge of us, Rosen argues, we are likely to be stereotyped and misjudged.[65]

What are the ethical responsibilities of those who possess personal information about clients or employees and are in a position to make critical decisions about services or jobs they might be seeking?

Customizing Web Sites

Another concern is that online tracking information may be used to tailor a Web site's presentation to a particular user. This would ordinarily not present a problem for an e-commerce site. Customization of user features might be considered a big time saver for someone who repeatedly visits an online store to buy books or make travel reservations. Why should they have to sift through all the other pages and registration forms when the site can take them straight to the services they use most often right when they enter the site?

This becomes a bit more complicated, however, when we consider news and political sites or the site of a company that has recently been the subject of controversy. Would it be ethical for the Web site of a political candidate to serve up messages tailored specifically to me based on my Web surfing habits—while at the same time suppressing viewpoints I might oppose?[66] What about a news organization that "pushes" entertainment and sports content at a user upon entering the site based on tracking data that shows that he or she regularly checks baseball scores and reads movie reviews? Many news sites do something similar when they allow users to customize the news they receive. But what if the site were to do this based on users' Web-surfing habits rather than some form they filled out? If news sites do this, are they abdicating their journalistic responsibility to deliver a comprehensive picture of the day's news to all readers? And isn't there a danger in assuming that a user always wants to see baseball scores and movie reviews first just because he or she visited those pages frequently in the past? What if, as mentioned above, the data in the profile were inaccurate?

Surreptitious Monitoring of E-mail and Web Use

Many Internet users browse the Web and use e-mail at work, often sending and receiving personal messages. However, these communications are not typically conducted in private, as employers have the legal right to monitor any communications over their phone and computer systems whether they tell workers or not. It appears as though many companies and government agencies are doing just that. Studies suggest that the Internet use of 14 million American workers is under continuous surveillance. Worldwide, that number is about 27 million, and sales of employee-monitoring software rose to about $140 million by 2000.[67] As a report from the Computer Professionals for Social Responsibility (CPSR) puts it, Internet "users should assume that they have no privacy in their workplace."[68]

Many employers defend surreptitious monitoring on the grounds that they have to keep tabs on workers who might be wasting time surfing Web sites unrelated to their jobs. They also claim that monitoring is necessary to catch workers who might be sending pornographic, racist or other harassing messages to co-workers. Employers fear that unless they monitor workers, and take swift action to discipline offending employees, they will face costly sexual harassment or racial discrimination lawsuits. Finally, they assert that monitoring communications systems helps them track workers who might be giving away company secrets.[69]

On the other side, numerous privacy advocates, union representatives and labor lawyers argue that continuous surveillance gives employers too much power to decide what constitutes privacy in the workplace. They maintain that workers are entitled to some measure of personal space in the office as well as the right to conduct a reasonable amount of personal communication. Unless there is specific evidence that an employee is breaking company rules, they reason, he or she should not be monitored. Another complaint is that while the messages of workers are being monitored, managers and executives are often excluded from surveillance, leaving them free to violate the same policies they expect workers to uphold.

Though employers are not legally required to adopt workplace privacy policies, the CPSR believes that basic ethical principles call for them to notify workers if they will be monitored and to tell them the reasons why such monitoring is being done. In addition, employers are asked to develop and clearly communicate policies regarding (1) the acceptable uses of e-mail and other company computer resources; (2) the enforcement of Internet rules, such as the intercepting or reading of e-mail or scanning of hard disks; (3) the penalties for noncompliance with these policies; and (4) the presence of any electronic monitoring systems that might be installed on workplace computers.[70]

Lurking

Internet communications can also be monitored secretly by one's peers through a behavior called "lurking." This is akin to eavesdropping, or perhaps listening to talk radio programs regularly without ever calling in. Internet lurkers typically log on to chat rooms and discussion forums or join e-mail lists and newsgroups with little or no intent of sending messages; rather they prefer to read the communications of others. But whereas the radio lurker cannot see the telephone numbers of those who call in, the online lurker usually has access to the e-mail addresses and screen names of anyone who sends a message or participates in a discussion.

There is nothing to prevent a lurker from copying a newsgroup or e-mail message from another subscriber and spreading it around the Internet, effectively making the original sender's e-mail address available to millions of other users. Likewise, lurkers in restricted chat rooms and forums may also be able to record messages and circulate them around the Net.

Lurking is practically impossible to prevent, though it does raise some ethical questions. For example, should journalists, academic researchers or public relations professionals gather information by lurking in chat rooms? If they do, what are their responsibilities with regards to the privacy of the other participants? Should journalists identify themselves accurately and state their purposes upon entering a chat room or logging on to a message board? Is it okay to lurk for a while before identifying oneself? Is it okay to quote from a message posted in a chat room?

Naturally, there will be differences of opinion about what is proper online behavior. Many chat rooms, e-mail lists and other Internet communities have general rules of etiquette that users are expected to follow. Journalists doing research online might find it helpful to review these guidelines before joining a chat room or signing up for an e-mail list. It is also worth noting that the practice of misrepresenting oneself while gathering information is generally discouraged. For example, the SPJ code of ethics states: "Journalists should . . . avoid undercover or other surreptitious methods of gathering information except when traditional open methods will not yield information vital to the public. Use of such methods should be explained as part of the story." The SPJ code also reminds journalists to "recognize that private people have a greater right to control information about themselves than do public officials and others who seek power, influence or attention."[71]

Traditional ethical standards for academic researchers are somewhat similar. Anthropologists and others doing ethnographic studies are strongly cautioned against observing subjects under false pretenses. Writing in a textbook for media researchers, journalism professor Susanna Hornig Priest argues that "although the group being studied may not need (or want) to know all the details of the ethnographer's research interests, they must know what the ethnographer is doing and be given the right to refuse."[72]

Case Study: Gathering Information from Chat Rooms

The Internet and the Web enable journalists to gather more information from more sources faster and with more flexibility. Researchers and reporters can now tap a wide array of sources that didn't exist prior to the Internet—Web sites, chat rooms, bulletin boards, online forums and e-mail lists—to harvest information about practically any subject imaginable.

But while online media provide easy access to information and diverse Internet communities, they present new ethical dilemmas. The medium can tempt journalists to conceal their own identity while trolling for information and "listening in" to the conversations of Web-based communities and online discussion groups. It also makes it much more difficult—perhaps even impossible—to verify that a person's identity is really what he or she claims it is—age, sexual orientation, ethnicity, gender and all the other fundamental ways we categorize people become invisible online.

Not only is verification difficult but serious questions about privacy also arise. For example, how should a journalist doing research on a particular group or organization approach a restricted Web site or chat room that requires members to register before gaining entrance? If a journalist did gain access to one of these sites either by registering as a reporter or through deceptive means, is it an invasion of privacy? From another angle, should journalists quote liberally from online interviews and chat sessions? Should they quote from them at all?

These questions and others come to the fore in the following case.

The Case: There's a Reporter in the Chat Room

In 1999, journalist and author Jennifer Egan was developing a story for *The New York Times Magazine* about gay and lesbian teenagers who were using the Internet to meet and interact with other homosexual teens.[73] As part of her research, Egan spent several weeks visiting Web sites geared to the gay community and lurking in several chat rooms that were restricted to gay and lesbian teens. Though Egan was in her 30s, she easily gained access to the chat rooms by entering a fake identity and age on the site's registration form.

After a period of observing communication among chat room participants, Egan posted a message identifying herself as a reporter. She also asked if anyone would like to participate in interviews about gay life online. Egan's initial request did not elicit much response, but in time she began receiving e-mail messages from teens who agreed to be interviewed for her story. Egan gradually learned that many of the teens were still living "straight" lives offline and that the Internet had become a kind of refuge for them—the only environment in which they could openly discuss their sexuality and interact with other gay teens. Some of these closeted kids were terrified that a family member or friend might uncover their online activities, so they created separate screen names and instant messaging services specifically for visiting gay chat rooms.

Eager to move beyond e-mail and Web chats, Egan began making telephone contact with some of her sources. In particular, she began having regular conversations with a 15-year-old boy who would become the primary subject of her story. During one conversation, Egan proposed the idea of traveling to his hometown in the rural South to conduct a face-to-face interview. The teenager agreed but warned Egan that they must be very careful to avoid being seen together. He feared that family or friends in his close-knit community would get suspicious if they saw him talking to an older stranger.

After the interview, Egan and the teen continued to communicate online, exchanging numerous e-mail messages. Egan also continued her correspondence with other gay and lesbian teens she had met in chat rooms. As the teens became more comfortable with her they began to discuss not just the loneliness of being a gay teen but also sexual activities they had engaged in, with both online and actual physical partners. Through these frank discussions, Egan learned that many of her sources had been pursued by much older adults who had been posing as teens in the chat rooms.

After more than a year of research and writing, Egan's story, "Lonely Gay Teen Seeking Same," appeared as the cover story in the December 10, 2001, issue of *The New York Times Magazine*. The 8,000-word article stressed the value of the Internet for early exploration of sexual identity, especially for teens who are isolated and worried about their parents' and classmates' reactions. The piece also mentioned the danger posed by adult sexual predators and the lack of security surrounding chat rooms and Web sites that claim to be only for teens. Egan brought many of these points to life by quoting at length from e-mail and instant messaging interviews with her sources. In several passages, she quoted teens discussing their emotional and sexual relationships with boyfriends and girlfriends, some of whom they had met or had sex with online. Egan did not reveal the identity of her subjects, referring to them only by their first names or the first letter of their first name.

Ethical dilemmas and rationale. As a reporter schooled in traditional methods of researching and interviewing, Egan says this story presented ethical challenges unlike any she had ever encountered. At each step, she says she tried to follow the basic ethical principle—do no harm—though it was often difficult to determine whether or not a particular action could cause harm. For example, when she began the story, Egan felt it was necessary to conceal her identity and age while registering to enter certain Web sites and chat rooms. Had she entered the correct information, Egan might have been denied access, thus losing the opportunity to locate and contact sources for her story. As a general rule, failing to identify one's self accurately is seen as deceptive and, therefore, would be a highly suspect action. But Egan may have provided a service to readers by exposing the weaknesses of chat room registration systems.

To complicate matters further, Egan lurked in several chat rooms before revealing her identity. This allowed her to get a feel for the normal flow of communication among participants, which she felt was essential to advancing her knowledge of gay online culture. Had she announced herself immediately, she might have disrupted the environment.

It's important to note that many people join "gated" online communities with the assumption that there will be some measure of privacy—that they will be able to protect themselves from prying eyes. This raises an important question: Are "members only" Web sites and chat rooms public or private places? Some journalists argue that they are public because, as Egan proved, it is relatively easy to circumvent registration procedures. Participants, they say, should assume that outsiders can and will eavesdrop on their discussions. In a sense, everything's public on the Net. However, online journalism critic J.D. Lasica frowns on reporters lurking in chat rooms and argues that it's ethical only "when the subject is of significant public importance." Lasica also indicates that using quotes from chat room discussions or bulletin boards without permission is "considered bad form" among avid Net users and may be unethical for journalists.[74] This would be especially true of comments that are of a deeply personal nature and those made by minors.

Egan's research was further challenged by the simple fact that she could not always be sure exactly whom she was observing or even interviewing. Internet communication allows one to conceal his or her identity, just as Egan initially did, and to construct an online persona that is far different from one's physical self, as many of her sources did. How, then, could Egan verify that her sources were, indeed, gay teenagers, and how could they be sure that she was a writer working on an article for *The New York Times Magazine*? Egan says when she did introduce herself in chat rooms, she clearly stated her occupation and her intention to conduct interviews with gay teens. She instructed those who wished to participate to e-mail her and promised not to reveal their identities. As for confirming the chat room's participants' identities, Egan says she paid close attention to the wording of each message she received and tried to verify the sex and age of key sources through follow-up telephone conversations. At one point, Egan reports, she stopped communicating with a source after sensing that he was much older than he claimed to be. As noted above, on only one occasion did she conduct a face-to-face interview with a source.

Lessons learned and suggested protocols. Looking back, Egan recognizes that her lack of exposure to the Internet played a factor in making some of these issues harder to resolve. As she put it, "technology changes the nature of communication," which forces writers to rethink traditional methods of conducting background research, doing interviews and choosing which statements to quote and which to leave out. To some extent, the design of the medium will dictate how it is used. But at the same time, the ethical journalist must try to minimize harm while using that medium. Conflicts require creative solutions guided by the journalist's sense of social responsibility. In many ways, Egan benefited from the fact that she wasn't under heavy deadline pressure—the story was more than a year in the making. This allowed her to slowly get a feel for online culture and to build a high level of trust with many of her sources. Other journalists will, no doubt, have the added pressure of much tighter deadlines. Lasica advises them to consult with colleagues who are more familiar with the medium and the norms of acceptable online behavior. If time permits, nonjournalists steeped in Internet culture can also be tapped for information.

Along with knowledge of the medium, Aly Colón, ethics faculty member at the Poynter Institute, suggests that each reporter ask him- or herself the four following questions before attempting to gather information from restricted Web sites and chat rooms:

- How do I plan to access the chat room or Web site?
- Will I identify myself as a reporter and state my intentions?
- Will I ask permission to quote participants or even to monitor their discussions?
- How will I authenticate what I read in the discussions?

In searching for answers to these questions Colón advises journalists to consider what they might do if they were in a comparable situation offline. Analogies from one medium to another can be difficult, but the exercise may be helpful in getting journalists to think more deeply about the privacy of others.

No matter what form the medium takes, Colón argues, journalists should address newsgathering challenges with respect for their "core values." Put another way, the settings in the online world might be different—virtual rather than physical—and the identities might not always be authentic, but behind those identities in those virtual spaces are real human beings. Decisions about how to observe and communicate with them ought to be guided by basic ethical principles, as should decisions about how much information about them will be revealed to a public they might be hiding from.

Case Study: Publishing
Public Records Online

We live in an age in which the law and technology have converged to make it possible for journalists to publish an unprecedented amount of information.

From the legal side, both the courts and legislative bodies have allowed vast stores of information, from birth records and bankruptcy filings to the addresses of criminals, to be placed in the public domain. Even when information of this sort was available in the pre-digital age, journalists were constrained by the limits of space (in print publications) and time (in broadcast media) from disseminating it to their audiences; they could summarize or analyze the data, but it was virtually impossible to publish all of it. Enter technology and the evolution of digital media, particularly the World Wide Web. Today, Internet publishing tools make it possible for journalists to efficiently digitize, publish and archive public records and databases. Gone are the days when the medium was an obstacle to publishing enormous streams of public information.

But advances in legal and technological freedom don't necessarily make the job of a journalist any easier. They may, in fact, make it much harder as reporters, editors and media managers face a whole new set of ethical responsibilities that did not exist with traditional media.

Consider the following case in which online journalists were confronted with the question of whether or not to publish the contents of a public database that included the names and addresses of convicted sex offenders.

The Sex Offender List:
To Publish or Not to Publish?

Following the lead of New Jersey, which enacted the famed Megan's Law in 1994, the Ohio Legislature in 1997 passed a law requiring convicted sex offenders to register with sheriff's departments in the counties in which they lived. State lawmakers also classified the registration lists, which included biographical information about the offenders, as public information.

In accordance with the state's decision, the sheriff's department in Cuyahoga County, home to the city of Cleveland, made its list of convicted sex offenders available to the public. But in order to obtain a copy, residents first had to make a personal visit to the sheriff's office and fill out an information request form. So while the information was available, some residents complained that it wasn't easily accessible.

Hoping to solve the problem of accessibility and promote a new Web site called NewsNet5, Cleveland TV station WEWS and its online partner, Internet Broadcasting Systems (IBS), obtained the Cuyahoga County list from a city councilman and decided to publish it online.[75] NewsNet5 executive producer Colleen Seitz says the information was "double-checked and triple-checked" for accuracy before the list was cleared for publication.

On June 10, 1998, newscasters on the WEWS evening broadcast advised viewers that they could see the list by logging on to the NewsNet5. On a web page headlined "Sexual Offenders Next Door: Find Out Who Lives in Your Neighborhood," the names and addresses of convicted sex offenders living in Cuyahoga County were preceded by the following message:

> When a convicted sexual offender moves into your neighborhood, you have a right to know about it—that's the law. But the law does not make it the government's responsibility to notify you. Instead you must make a request for those names, and then make arrangements to pick up the information. NewsNet5 and NewsChannel5 obtained a list of convicted sexual offenders in Cuyahoga County from the office of Ward Four Councilman Dean DePiero. ZIP codes were NOT available. These are the last known addresses.

Information about those classified as "Habitual Offenders" was displayed first followed by the names and addresses of those classified as

"Sexual Predators." Finally, the names and addresses of those convicted of "Sexually Oriented" offenses were listed. Brief definitions of the three classifications were included with the list and a disclaimer at the bottom of the Web page cited the Cuyahoga County Sheriff's Department as the source of the information.

Ethics specialists studying online journalism agree that Web sites can provide a valuable service by publishing public records. However, they caution editors and reporters to avoid being tempted by the medium into publishing first and asking critical questions later. Colón of the Poynter Institute suggests that journalists consider several questions before making a final decision about whether or not to publish a particular database:

• How does the information fit into the particular story being told?
• What is the journalistic value of the information—does it add meaning to the story?
• Does the information help the public make decisions?
• Can harm be caused as a result of the information being published?[76]

In addition to these questions, the Poynter Institute suggests that the following guidelines be followed once a decision has been made to publish databases:

• Have a mechanism and a commitment for keeping the data updated and accurate. When it cannot be revised as necessary and when there is a potential for harm caused by old data, the database should be removed in a timely manner.
• Ensure that the data are evaluated as thoroughly as possible prior to publication, using traditional high standards of accuracy, reliability, validity and credibility.
• Reveal the authorship/ownership, scope, validity, and limitations of the data made available to the public.

The NewsNet5 rationale. Colleen Seitz recalls that making the decision to publish the sex offender list was not an easy task. Several staff members raised serious objections about the widespread dissemination of the list, which they believed was an invasion of the convicts' privacy. But, on the whole, Seitz says the majority of the staff believed that the list should be published. As the manager of the Web site, Seitz worked closely with the news director and assistant news director at WEWS and managers at IBS. This is not an unusual arrangement for Web sites affiliated with local TV stations or newspapers—many of them are developed and maintained through partnerships with other news outlets or companies, such as IBS, which specialize in providing online development and hosting services to dozens of TV stations. Thus, people with experience in both traditional and online media were involved in making the final decision.

Ultimately, the managers decided that the public should have access to the names and addresses of convicted sex offenders without having to make personal trips to the sheriff's department. As Seitz put it: "It always came back to the fact that this was public information . . . we weren't giving them [the public] anything they couldn't get themselves, we just made it more accessible." In short, Seitz and the other managers felt that they were performing a public service by publishing the information online—the public's right to know outweighed privacy concerns. Jay Maxwell, an editor with IBS at the time, echoed Seitz's comments in a memo to company employees: "I believe that these kinds of databases —controversial or not—point to the strength of our brand of journalism."

The reaction. The sex offender list quickly became a popular feature on NewsNet5 and staff members reported overwhelmingly positive feedback from visitors. However, there was opposition from some in the community who voiced concerns about privacy. Family members of some of the offenders raised objections, arguing that the information put them at risk for harassment and retaliation.

News managers also learned that some offenders on the list had moved and failed to report their new addresses. Concern grew that keeping old addresses online would put others at risk of being labeled as sex offenders or, perhaps, of being attacked by vigilantes. The concerns were not unfounded. In 1999, a mentally retarded Texas man was attacked and beaten by four men who thought he was a sex offender. The man lived in the same group home in which a convicted sex offender had once lived. The offender had moved away several months earlier, but the address of the home was still listed in the state's Internet database. After the beating was reported, the address was removed.

Despite the criticism and concerns about the accuracy of the list, the NewsNet5 staff chose not to remove it. Instead, they decided that they would keep the information online and update it once a year. The staff also started adding lists from other counties in the Cleveland area. Seitz says each new list is checked for accuracy before it is published. However, not all lists have included the date of publication (or latest update) and the source of the information.

This could be a critical lapse according to Barb Palser, an online news editor and contributor to *American Journalism Review*. She recommends that Web journalists always include the date of publication and the source of all documents and public records.[77] To do less would be to invite confusion and, perhaps, be guilty of misleading the audience. Additionally, Web journalists—and Web page designers working for news organizations—are discouraged from scattering pieces of data from a particular list or database across numerous pages on the Web site. Breaking up the data may add flexibility, but it is also likely to cause problems for a busy Web editor who is trying to update the records—it increases the chances that old or inaccurate information will be left online.

As mentioned earlier, legal and technological advances can liberate journalists from old constraints. However it has also been argued that each increase in freedom brings with it a proportionate increase in responsibility. If the Internet has, indeed, increased journalistic freedom, then it would follow that online journalists have an increased obligation to use the technology responsibly. In short, ethical decision making requires the journalist to think beyond what he or she *can* do and contemplate what he or she *should* do.

Complying with Governmental Demands

Finally, online professionals may be confronted with the question of how to respond to demands from government agencies to turn over customer information or to allow law enforcement to install surveillance programs. Telecommunications workers fear that extensive cooperation with law enforcement would make them a de facto arm of the police in addition to possibly compromising the privacy of innocent customers. From another position, workers and their employers often sympathize with police efforts to catch criminals and don't want to withhold information or block surveillance efforts that might prevent future crimes and save lives.

This is not a new dilemma. Major ISPs have reportedly been fielding hundreds of requests each year from police to turn over information relevant to criminal investigations and lawsuits. But with the development of sweeping new surveillance systems such as Carnivore and the passage of the USA PATRIOT Act, ISPs may be more likely to be asked to help police and government security agencies track and intercept messages, and they are likely to find it harder to refuse such requests.

In the days immediately following September 11, 2001, major Internet services AOL and Earthlink as well as some smaller ISPs were approached by law enforcement and asked to cooperate with terrorist investigations. The operator of an online service in Sacramento, Calif., told *The New York Times* he acted personally to make sure law enforcement got the customer information they were looking for.[78]

As mentioned previously, libraries are subject to government requests for information about suspected patrons, including lists of books they've checked out and the names of Web sites they've visited. In order to protect the privacy and intellectual freedom of patrons, many libraries have historically kept very few records on individuals' reading or viewing history. For example, when a patron returns library items on time, they are usually expunged immediately from the patron's records. Likewise, many public libraries configure patron Internet browsers so that Web sites visited are not recorded in a history file.

With new powers, though, government agents can approach libraries and ask them to install Internet monitoring software that would override existing privacy protections. In addition, librarians would be forbidden from telling anyone that patrons are being monitored or that records have been requested.

Naturally, this can put librarians in a tight spot between the rights of their patrons and the demands of law enforcement. The American Library Association (ALA) has reminded members that it is their ethical responsibility to seriously question federal search warrants or information requests before taking action. In June 2002, the ALA also issued a statement instructing librarians to do all they can to protect patron records, concluding, "rights of privacy are necessary for intellectual freedom and are fundamental to the ethics and practice of librarianship."[79]

Summary

Privacy and the desire for protection against unwanted intrusions into a private zone is often considered a fundamental human right and has deep historical roots. Two types of privacy that are of particular concern to media professionals are those termed *information privacy* and *communication privacy*.

Although there are treaties and universal declarations that uphold the right to privacy, the United States does not have a general privacy law nor is there any explicit privacy protection spelled out in the Constitution. However, over the years, the Supreme Court ultimately carved out a derived right of privacy by drawing from existing amendments in the Bill of Rights.

The evolution of privacy law and the legal levels of protection have been greatly influenced over time by technology. Among the technologies that created new concerns over privacy were the telegraph and camera and eventually the telephone and computers. The earliest of these new forms of intrusions ultimately led attorney Louis Brandeis to proclaim in 1890 that an individual had a "right to be let alone." Brandeis would later serve on the Supreme Court, where he continued to champion privacy rights and other civil liberties.

The development of the computer, and then the Internet, caused fresh privacy concerns. One of the earliest worries was that the government and private companies both had the ability to collect personal data on millions of individuals and to create massive databases of personal information. Another related to electronic tracking and surveillance by corporations and the government. Companies can track and monitor the habits and purchases of current and potential customers while they are surfing the Web by placing a small file on the user's computer, called a cookie, which tracks the user's online activities. Consumer protection and privacy advocacy groups have lobbied Congress to try to restrict and place limits on this kind of tracking and have worked to try to force companies to obtain consent from the user as to how his or her personal information may be utilized.

Another new type of surveillance occurs when firms install tracking software on their employees' PCs in order to monitor their Internet use habits while on the job. There are also newer privacy concerns regarding the U.S. government and Internet surveillance. Recent federal initiatives to combat terrorism, such as the passing of the USA PATRIOT Act, allowed for increased surveillance and a lower level of privacy protection overall, and have increased concerns regarding a diminishment of the Internet user's privacy.

Privacy advocates are also concerned about credit services and employers using inaccurate data to make unwarranted negative decisions on individuals, as well as the potential for people to make incomplete and unfair judgments about others based on out-of-context comments made in a chat room, their Web browsing habits, or other Internet usage.

Certain privacy issues raise ethical concerns particularly for journalists. For instance, is it appropriate to use comments made in Internet discussion groups as source material? Using a chat room participant's remarks raises questions regarding whether and how a reporter needs to identify his- or herself when entering that discussion group and what kinds of consent must be provided by a participant in order to use his or her comments.

Another privacy issue of concern for online journalists is whether to link to or post existing public records. Doing so will make it much simpler for the public to find out personal, even damning, information about others in their community, which may range from divorce cases and civil judgments to sex offender listings. The data is already public, but there are considerations that need to be taken into account before deciding to make the information so widely and easily available.

Critical Thinking Questions

- In order to better understand the reading habits of the readers of your online news site, you are considering adding software to track which users visit which sections and in what order. Are you ethically required to let your readers know about this tracking software? Explain.
- You require users of your online news site to identify themselves upon registration with some basic identifying data. What are you ethically permitted to do with that data: Rent it to other marketers? Use it for your own marketing purposes? Use it to study the demographics of your readership?
- What would an "ideal" privacy policy look like for an online media's Web site? Discuss the ethics of the opt-out vs. the opt-in option.
- Can you think of a circumstance where it would be ethically justifiable for you to conceal your identity as a reporter when entering an online discussion group? If so, what would be the larger ethical principle that made doing so permissible?
- When you are sitting at a local restaurant, do you feel that your conversations are public or private? How about at a club that you are a member of? What about when you are chatting online? How would you feel if you discovered that a journalist was "listening" to

you and your friend's conversations at your local diner? How
would you feel if you discovered that a journalist was monitoring
your conversations on an online discussion group? Would you feel
differently? Why or why not?
- How would you feel if something you posted in an online discussion group turned up in a feature story some months later, quoting
you anonymously?
- In making a determination as to whether or not to post public civil
and criminal records on your Web site, what is the highest journalistic principle that should guide your decision? If you chose to post
the records, would you have an obligation that the information is
accurate and up to date? If so, how would you ensure this?

Key Terms

Carnivore
Children's Online Privacy Protection Act (COPRA)
Cookies
Fair information practices (FIPs)
Information privacy
Lurking
Opt-in/Opt-out
Right to be let alone
Universal Declaration of Human Rights
USA PATRIOT Act
Web bugs

Recommended Resources

Books

Alderman, Ellen, and Caroline Kennedy. *The Right to Privacy* (New York:
Knopf, 1995).
Garfinkel, Simson. Database Nation: The Death of Privacy in the 21st Century
(Sebastopol, CA: O'Reilly, 2001—paperback).
Rosen, Jeffrey. *The Unwanted Gaze* (New York: Random House, 2000).

Associations and Institutes

Berkman Center for Internet and Society. Web site: http://cyber.law.harvard.edu/
Center for Democracy and Technology. Web site: http://www.cdt.org/.
Computer Professionals for Social Responsibility. Web site: www.cpsr.org.
Electronic Privacy Information Center. Web site: www.epic.org.
Federal Trade Commission. Web site: www.ftc.gov.
Privacy Foundation. Web site: www.privacyfoundation.org.
Privacy International. Web site: www.privacyinternational.org/.

CHAPTER 5

Speech: "The Indispensable Condition"

Chapter Goals:

- Trace the historical roots of an ethic for free speech.
- Examine the birth of the First Amendment and its evolution via seminal Supreme Court rulings.
- Identify the major exceptions to the protections of the First Amendment.
- Explore ethical dilemmas facing online journalists arising from the clash between the ethic of free speech and the Internet: permitting hate speech, posting and linking to offensive content, and complying with governmental demands,

The right to speak and express oneself freely and the right to disseminate information, ideas and opinions through the media without being censored are considered by many to be the most sacrosanct of human rights, deserving of full legal protection. As U.S. Supreme Court Justice Benjamin Cardozo wrote in the 1930s, freedom of expression is "the matrix, the indispensable condition, of nearly every other form of freedom."[1]

In addition to being fundamental to individual liberty and autonomy, legal and political scholars have also argued that free speech and a free press are necessary preconditions for democratic governance. From this view, open, political debate and the uncensored dissemination of information about government is the method by which the citizens restrain tyranny and protect their collective sovereignty. Alexander Meikeljohn, a political philosopher and educator, was the most noted proponent of this view. "It is the mutilation of the thinking process of the community," he wrote in 1948, "against which the First Amendment to the Constitution is directed. The principle of the freedom of speech springs from the necessities of the program of self-government."[2]

Another view holds that free expression is essential to the search for truth and should be protected for intellectual and moral reasons. The 19th-century English philosopher John Stuart Mill believed that truth was best uncovered through a competitive exchange of ideas. In his seminal political essay *On Liberty*, he presented three reasons why unorthodox opinions should not be censored: (1) The censored idea might be true and received opinion might be false; (2) truth needs to be tested, otherwise it becomes "dead dogma"; and (3) there is likely to be some truth in all opinions—the free exchange of ideas will allow the most truthful parts to surface while also giving listeners, or receivers, a chance to identify and reject the false concepts.[3]

Recognizing and Protecting Free Speech

International Law

Generally speaking, free expression is protected on a global scale by numerous treaties and agreements that form the backbone of international law. Article 19 of the Universal Declaration of Human Rights (UD), for example, states:

> Everyone has the right to freedom of opinion and expression; this right includes freedom to hold opinions without interference and to seek, receive and impart information and ideas through any media and regardless of frontiers.[4]

Though the UD was adopted in 1948, two ideas contained in it are particularly relevant to the issue of Internet speech. First, the use of the words "regardless of frontiers" seems to indicate that a high premium is placed on allowing content to flow across political borders. Secondly, the phrase "seek, receive and impart information" is well suited to a medium that enables the user to be both reader/viewer and publisher. The International Covenant on Civil and Political Rights (ICCPR), adopted in 1976, uses virtually the same language as the UD but also allows that speech may be restricted when it infringes on "the rights and reputations of others" and when it threatens "national security . . . public order . . . or public health or morals." In addition to these qualifications, Article 20 of the ICCPR places an outright ban on "propaganda for war" and "advocacy of national, racial or religious hatred that constitutes incitement to discrimination, hostility or violence."[5]

Thus it appears as though international law supports the ability of people to communicate freely across borders through any media with limits on certain forms of libel, hate speech, obscenity and misinformation designed to promote war.

National laws. Though countries can be punished for violating international law, political and civil rights provisions are difficult to enforce. In a sense, sovereign nations are encouraged, even pressured, to protect speech in a manner that is consistent with the UD and ICCPR, but they essentially retain their right to craft their own constitutional provisions or legal statutes for protecting speech and the press and to devise mechanisms for enforcing them. As such, speech and press protections vary from country to country depending on the political and cultural context in which they are created.

For example, in making the transition from apartheid to constitutional democracy, South Africa adopted a bill of rights that is very much in line with international law. It protects the right of free expression for all individuals and the press, with exceptions for propaganda for war; incitement of imminent violence; and advocacy of hatred based on race, religion or gender.[6]

Similarly, in many European countries, the ugly experience of fascism and the Holocaust have caused some governments to ban racial or ethnic hate speech and literature as well as the public sale of Nazi memorabilia. Germany and France, for example, have placed severe restrictions on the sale of the book *Mein Kampf*, which they consider to be an incitement to racial and ethnic hatred and violence.[7]

In the United States, free speech and the press are protected by the First Amendment, which, of course, was ratified long before international civil rights covenants were adopted. It is also more libertarian than the South African bill of rights, as well as most European speech provisions. Before moving on to some of the specific ethical dilemmas that arise when communication goes global via the Internet, it might be helpful to review the First Amendment and the way it's been applied to the Internet. In doing so, we'll see where it differs from international law.

The First Amendment

"Congress shall make no law . . . abridging the freedom of speech, or of the press . . ."

—First Amendment to the Constitution
of the United States of America

The adoption of the First Amendment in 1791 was a watershed event in the long struggle for individual liberty and democracy. Never before had officers of government placed such a strong check on their own power. No doubt this precedent has had enormous influence on citizens in other countries struggling to establish constitutional democracies as well as those engaged in drafting international law. Looking back on its contribution to law, politics and communications, contemporary scholars have referred to the First Amendment as the "crown jewel" of American democracy.[8]

It is worth pointing out, though, that the succinct language of the First Amendment is distinctively negative; it simply says that Congress cannot make laws abridging free speech and the free press. It says nothing about censorship from nongovernmental bodies, such as corporations, or in semipublic places such as shopping malls. The Supreme Court, in fact, has ruled that the First Amendment does not apply to speech on private property; therefore, journalists working for a privately owned media company have no First Amendment protection unless ceded by their employers.[9]

The narrowness of the First Amendment is due primarily to the political context in which it was written. In the late 18th century corporations had very little political power and influence compared to today. Framers of the Bill of Rights were more concerned with balancing the powers of the strong federal government outlined in the new Constitution. James Madison, in particular, worried that a dominant political faction or party would use their legislative power to censor opposing views; therefore, the Bill of Rights should free speech and the press from congressional interference. Madison could not have anticipated that institutions outside the orbit of government would someday have the power to regulate speech.

Sedition: Congress makes a law. Despite its clear prohibition of government censorship, the First Amendment was circumvented by federal legislators numerous times during war and national crisis. In 1798,

the Federalist-controlled Congress attempted to silence critics by passing the Sedition Act, which outlawed "false, scandalous and malicious"
communication against the government.[10] This statute was partly aimed
at the Republican press, which supported Madison's and Thomas Jef-
ferson's political views.[11] Though the act expired in 1801, it wasn't until 1964, in *New York Times v. Sullivan,* that the Supreme Court declared
it to be unconstitutional.[12]

In 1917 as the United States entered World War I, Congress banned
the use of the mail to disseminate material deemed to be treasonous.
This effectively banned newspapers and magazines, such as *The Nation,* which published anti-war articles. Also during the war, Congress
passed the Espionage Act, which outlawed the publication or distribution of material that advocated resistance to the military draft as well as
other content judged to be useful to the enemy. Hundreds were convicted under the statute and were either fined or imprisoned. In a few
notable cases that reached the Supreme Court, justices upheld convictions under the Espionage Act on the grounds that during war it may be
appropriate to ban certain kinds of speech that are protected during
times of peace.[13]

Outside of wartime, the high court has tended to treat the First
Amendment much more literally, casting a critical eye on attempts by
government to restrict individual expression or access to information
and allowing that the concept of free speech is not to protect "free
thought for those who agree with us, but freedom for the thought that we
hate."[14] Since the 1960s, the Supreme Court has overturned most of the
federal laws designed to punish seditious political speech, giving Americans unprecedented formal protection for free expression.

The high court has also ruled that the First Amendment applies to the
states and their public units—schools, libraries and city governments—
as well as to the federal government.[15] In keeping with this principle,
states have the power to adopt their own speech protections so long as
they are equal to or greater than those of the First Amendment. Some
states have, in fact, gone beyond the Supreme Court to protect political
canvassing and petitioning in privately owned shopping centers.[16]

Unprotected Speech

Despite the court's liberalism in applying the First Amendment in the
latter half of the 20th century, there are several types of expression that

have historically been unprotected: obscenity, incitement to violence and fighting words, and libel or defamation.

Obscenity. According to the standard established in a 1973 Supreme Court case, a work is obscene if a judge or jury determines that (1) the average person, applying contemporary community standards, would find that the work appeals to the prurient interest, (2) the work depicts or describes sexual content in a patently offensive way, and (3) the work lacks serious literary, artistic, political or scientific value.[17] In deciding all three parts of the Miller test, as it's called in reference to the name of the case, the work must be taken as a whole; and if the work meets each element, it is not protected by the First Amendment.

In allowing for "community standards" to be used as a guide to decide the first part of the test, the Court assumed that jurors selected from the community in which the work in question was displayed or published would reflect local attitudes and thus be able to judge if the work was prurient—that is, capable of arousing sexual desire.

It is important to note that, in a legal sense, obscenity is not a synonym for nudity, indecency or pornography, which are protected so long as they don't violate the Miller test. Child pornography is an obvious exception, though, in that the courts have ruled that it constitutes a form of child abuse, thus states are entitled to outlaw the production, dissemination and possession of it.[18] In other words, Americans have the constitutional right to offend but not to produce obscene works or to participate in child abuse.

Incitement to violence and fighting words. Incitement to violence refers to expression directed against the government that is judged to be "directed to inciting or producing imminent lawless action" and is "likely to incite or produce such action."[19] This standard comes from a 1969 case in which the Supreme Court overturned the conviction of a Ku Klux Klan leader who made a speech to other Klan members in which he threatened "revengance" against the federal government for its support of civil rights legislation. In making their decision, the justices argued that the Klan leader's remarks advocated the idea of taking violent action against the government rather than inciting actual violence that was likely to occur immediately after he made his remarks.

Fighting words pertain to individual confrontation rather than threats against government. The Supreme Court has ruled that words spoken by one person to another in a face-to-face encounter that are likely to incite an immediate breach of the peace are not protected by the First Amendment.[20] Put simply, if the speech in question occurs between two people and is likely to provoke immediate physical violence, it may be abridged.

Though some state and local governments have tried to criminalize racial, ethnic or gender-based hate speech on the grounds that the words themselves are capable of causing emotional wounds, the courts have considered such laws to be unconstitutional. Thus the First Amendment protects hate speech, unless the speech is likely to incite immediate violence or constitutes a threat of violence directed at a specific person, whereas international civil rights covenants and the national laws discussed previously generally do not.

Libel/defamation. This category refers to the communication—words or images that are published or broadcast—of falsehoods or misleading information that damages "a person's reputation, standing in the community, or ability to associate with others."[21]

In the landmark *New York Times v. Sullivan* case, the Supreme Court for the first time applied the First Amendment to a question of libel, which until then had been left up to the states to regulate. The justices ruled that a public official could not be awarded damages for publication or broadcast of "a defamatory falsehood relating to his official conduct unless he proves that the statement was made with 'actual malice'—that is, knowledge that it was false or with reckless disregard of whether it was false or not."[22] So not only did the information have to be false, the defendant also had to prove that the communicator intended to cause harm by deliberately lying, or that he or she showed very little concern for verifying the information —a tough, though not impossible, standard for a public official to meet. In making their test, justices argued that democracy required vigorous criticism of government officials even if it contained factual errors presented unintentionally. Naturally, members of the media as well as dissident political groups and other free speech advocates praised the court's decision.

A series of subsequent cases expanded the *Times-Sullivan* test by applying it to "public figures" (those not affiliated with government who are in the public spotlight such as coaches, athletes and entertainers) and "private matters" relevant to a public official or candidate's fitness for office. However, the Supreme Court backtracked in *Gertz v. Welch*,[23] another landmark case in which justices drew a distinction between public officials and public figures vs. private individuals that has remained largely intact. Those in the former categories are considered to have better means, such as access to major media outlets, to counter false information. Private citizens, on the other hand, are considered less able to refute untruths; therefore, they must at least prove negligence, and not necessarily actual malice, even if the information in question deals with a public issue.

Special Categories

Restrictions on commercial advertising that are designed to protect against consumer fraud are also constitutional, as are those that regulate the "time, place, or manner" in which speech is carried out.

On this last point, rules are permitted only if they "are content-neutral, are narrowly tailored to serve a significant government interest, and leave open ample alternative channels of communication."[24] An example would be rules governing the issuance of parade permits or rules limiting noise levels near hospitals or in residential neighborhoods during overnight hours.[25]

Regulations governing access to the broadcast airwaves also fall into this last category, though at first glance, they appear to clearly violate the First Amendment. However, unlike printing presses, which are considered to be private property, the airwaves are public resources and are technologically scarce—there are far more people wanting to broadcast than there are frequencies to go around. As such, the Supreme Court has ruled that the FCC has the authority to regulate the medium by allocating frequencies in a manner that promotes the "widest dissemination of information from diverse and antagonistic sources."[26]

The FCC is also empowered to establish and enforce limits on "indecent" programming by designating certain hours of the day or night during which such programs can run.[27] These rules don't apply to cable TV operators who have the right to determine when questionable content will run. The rationale for the difference between broadcasting and cable is that the former is far more accessible to children because it is so pervasive—it can be accessed over the air for free. Basic cable services as well as premium channels such as HBO, on the other hand, are typically available only to subscribers, giving parents more control over what their children see and hear.

It is worth lingering on this last point a bit longer. In making its decisions, the Supreme Court has recognized that "each medium of expression presents special First Amendment problems". In doing so, they have not applied the First Amendment equally across media platforms; rather they have tried to address each medium according to its technological structure. One can easily reject offensive print or cable TV content, it is reasoned, by not paying for it—by not bringing it into the home. An offensive broadcast, on the other hand, is much tougher to avoid. Justice John Paul Stevens has written that such content, in the case of television, is transported into the privacy of the home where it "confronts the citizen," intruding on his or her "right to be left alone."[28] By alluding to Justice Brandeis' famous quote, Stevens suggested that personal privacy trumps the First Amendment rights of the broadcaster when content is deemed offensive.

The First Amendment, International Law and the Internet

Indecency, Obscenity and Government Regulations

When considering regulations on Internet speech, U.S. courts have generally applied Justice Stevens' logic. That is, they have treated Internet content more like print or cable TV content based on the belief that Internet users must actively retrieve, or pull, content, whereas broadcasting "pushes" content toward the viewer or listener.

The first attempt by Congress to censor the Internet illustrates this point quite clearly. In 1996, lawmakers passed the Communications Decency Act (CDA), which called for criminalization of the "knowing" transmission of "obscene or indecent" messages and the sending or displaying of "patently offensive" content through the Internet to anyone under 18 years of age. A federal court ruled that the act was unconstitutional in part because it awkwardly lumped obscene, indecent and offensive content into one broad category. The court also held that the CDA would have applied one obscenity standard to all communities and prevented adults from accessing Internet content they would be able to access legally through traditional media. In the ruling, Judge Stewart Dalzell made a special point to contrast the technological structure of the Internet with the more invasive, one-way nature of broadcast media, writing that "the Internet may fairly be regarded as a never-ending worldwide conversation." He then argued that as the "most participatory" medium "yet developed," the Internet "deserves the highest protection from governmental intrusion."[29] Regulation in the form of the CDA, asserted Dalzell, would raise barriers to entry, thus defeating the Net's basic structure and chilling free speech.

In 1997, the Supreme Court, with Justice Stevens delivering the opinion, upheld the district court's decision and noted that while the government had a strong interest in protecting children from harmful content, the CDA would end up suppressing a wide range of legal speech. Justices also ruled that regulations similar to those governing broadcast content were not applicable to the Internet. The medium is neither scarce nor "invasive" and as such, it deserves the fullest First Amendment protection, according to the Court.[30]

Undaunted by the failure of the CDA, Congress enacted the Child Online Protection Act (COPA) in 1998. However, this time they narrowed the focus to prohibit commercial communication on the World Wide Web that is "harmful to minors" under the three-part Miller test. Despite the revisions, the act was quickly challenged and a federal court issued a preliminary injunction that barred the government from enforcing the act on the grounds that it was unlikely to survive "strict scrutiny." In May 2002, after the injunction was affirmed by a federal appeals court, the Supreme Court also upheld the injunction but sent the case back to the appeals court, ordering it to decide several constitutional questions. In March 2003, the appeals court ruled that the COPA violates the First Amendment because it improperly restricts access to a large amount of online content that is lawful for adults.[31]

The courts have also addressed two other statutes that would regulate obscene or indecent content on the Internet. In April 2002, the Supreme Court struck down the Child Pornography Protection Act (CPPA), which banned "virtual" child pornography—that is, sexually explicit material that appears to involve a minor such as animations, digitized paintings or films and TV shows in which an adult plays the part of a minor. In striking the statute, the court ruled that the law unconstitutionally prohibited "the visual depiction" of the idea of teenagers engaged in sexual activity rather than actual child pornography, which is already illegal. This, according to Justice Anthony Kennedy, would be tantamount to the government becoming thought police.[32]

The other law, called the Children's Internet Protection Act (CIPA), was declared unconstitutional in May 2002 by a federal court and, as of late 2002, was on appeal before the Supreme Court. The act ordered federal agencies to withhold funding from public libraries that failed or refused to install filtering software on computers with Internet access. The court wrote, "CIPA requires them [libraries] to violate the First Amendment rights of their patrons, and accordingly is facially invalid."[33] It also held that libraries should have the option of adopting less restrictive means, which are already in use in many libraries, to protect children from pornography such as (1) offering families a choice of filtered or unfiltered access for their children, (2) education and Internet training courses, (3) adoption and enforcement of Internet use policies by library staff, and (4) strategic placement of unfiltered terminals and/or the use of privacy screens to limit viewing to adult patrons.

Some critics argue that commercialization of the Web, media consolidation and new copyright laws (see chapter 6) are already transforming the Net into a broadcast-style medium. To give one example, download speeds on cable broadband connections and digital subscriber line (DSL) services are many times faster than upload speeds, creating an uneven publishing field between major content providers and consumers. In addition, offensive content, such as pornographic spam, is increasingly being "pushed" toward the user without consent. Despite these trends, the dominant judicial opinion is that the Net affords far more user choice than traditional media and retains enough of its original open structure to facilitate what Judge Dalzell called "a never-ending worldwide conversation."

Before moving on, it is important to note that one legal principle expressed in the early anti-obscenity measures has been accepted by the courts: the notion that individual Internet users or Web publishers, rather ISPs or Web hosting services, are responsible for illegal content. The exception would be cases in which the ISP or Web service generates the content. A somewhat similar principle holds for libel cases; individual users are responsible for libelous comments posted to Web forums and chat rooms. In cases where the operator of the service edits the messages or moderates the discussion, they assume greater responsibility.

Political Protest, Hate Speech and Border Disputes

While the Internet conversation does include pornographic messages, it also includes a wide array of educational content, as well as lively political dialogue, criticism and commentary. In this respect, the Net has created new opportunities for expression for emerging democratic movements in totalitarian states. In Iran, for example, political reformers and independent journalists are working around strict press censorship laws and state-run broadcasting services by publishing critical content on the Web. As one Iranian online journalist told a Reuters correspondent, "Technology always wins, and therefore the closure of reformist newspapers is useless when there is the Internet."[34]

Clearly international law protects vigorous political criticism so long as it doesn't include "propaganda for war" or "hate speech" that incites "discrimination, hostility, or violence." But herein lies a problem. International law is difficult to enforce. Some governments have not signed international civil rights covenants and refuse to accept the legitimacy of international law. Among those that do, perceptions of what is or isn't hate speech or propaganda may differ from country to country, even citizen to citizen—one person's political commentary is another's hate speech, just as one's art is another's pornography. So while the Internet affords great opportunity for democratic dialogue, many civil libertarians worry that broadly interpreted and strictly enforced hate speech restrictions could suppress legitimate political discourse and independent journalism.

At the same time, we must remember that hate speech is protected by the First Amendment, and while Congress moved quickly to pass restrictions on pornography and indecent sexual content on the Internet, it has not pursued hate speech with similar zeal. Most of the attempts to censor hate speech online have come instead from schools, particularly high schools and colleges, and commercial ISPs, portals, and Web hosting companies, some of which have adopted content guidelines that customers using chat forums or publishing home pages are obliged to follow.

Self-censorship has not, however, kept U.S. online services free from conflict with the hate speech laws of foreign governments. In 2000, a French judge ruled that Internet portal Yahoo! had to block French citizens from accessing online auctions of Nazi memorabilia, which is restricted under France's hate speech laws. Yahoo! complied with the judge's order and removed the items, but it also asked a U.S. court to decide whether or not French laws could be enforced on an American-based company. The court agreed with Yahoo! and ruled that the First Amendment trumped French law, making it unenforceable on U.S. companies. Not to be deterred, French groups that found the content offensive have charged Yahoo! and its CEO with violating the country's war crimes laws even though Yahoo! says it won't allow any such items to be posted again.[35]

The Global Internet Liberty Campaign (GILC), a coalition of civil liberties and international human rights groups, supports the application of international law to Internet speech. They also urge governments to "prohibit prior censorship of on-line communication" in any restrictive measures they do adopt, arguing, "The unique qualities of the Internet justify according the strongest protection to free expression on-line and should prompt a new vision of the right to receive and impart information 'regardless of frontiers.'"[36]

Despite the efforts of the GILC and other groups, followers of Internet law expect more countries to adopt and enforce hate speech laws that go beyond international law, which would further complicate issues of jurisdiction as well as cultural standards. Technology reporter Lisa Bowman highlights some of the difficult questions that confusing judicial guidance and inconsistent statutes bring to the fore: "Is putting up a Web site enough to warrant prosecution? Or must you target it to specific populations? What about disclaimers? Is it a deterrent to plaster a warning across your site saying 'these pages are for U.S. residents only?'"[37]

Even if courts ruled that online services must have a physical presence in countries with hate speech laws in order to be punished by them, some civil libertarians worry that the mere presence of such statutes and the threat of enforcement across borders would cause companies to further censor themselves, thus chilling controversial or unpopular speech throughout the Internet. In late 2002, researchers at Harvard University released a study indicating that the popular search engine Google had removed over 100 Web sites, most of them featuring anti-Semitic and white supremacist content, from its German and French listings. Google officials said they removed sites that had generated complaints in order to avoid legal liability.[38]

Libel Jurisdiction

Jurisdictional disputes have also arisen in Internet libel cases. The most notable example is *Dow Jones v. Gutnick*, a case in which a prominent Australian businessman, Joseph Gutnick, claims he was defamed by an article in the online edition of *Barron's* magazine. The case is especially noteworthy because it addresses the question of when and where publication legally occurs in cyberspace. The courts had to decide whether an article is published when it is uploaded to a server (the location of the server being the place of publication) or when it is viewed by a reader (the location of the reader being the place of publication). The Dow Jones Company, publisher of *Barron's*, argued that the case should be heard in the United States—New Jersey to be more specific—where its Web servers are located. Gutnick's attorneys sued in Australia, claiming that defamation occurred there when Australian citizens read the article. A lower court in Australia ruled in Gutnick's favor, arguing that publication occurs where and when the contents of an article are seen by a reader—where and when the alleged defamation occurs.[39] In December 2002 the High Court of Australia upheld the lower court's decision.

Critics of the lower court's argument, including many prominent international news organizations that joined Dow Jones in the case, argue that it opens publishers to libel lawsuits anywhere their content is downloaded, effectively placing them in the impossible position of having to conform to every libel standard in the world and perhaps every obscenity or hate speech standard as well. Michael Overing, a professor of Internet law and columnist for the *Online Journalism Review*, agrees with this view and suggests that courts in other nations may follow the Australian lower courts' logic when confronted with similar lawsuits. "What we thought we knew about jurisdiction," Overing writes, "is now falling apart in cyberspace."[40]

Others suggest that national governments, under pressure from news organizations, may attempt to work out libel treaties or agreements similar to those hammered out over global copyright issue.[41] However, reaching broad international consensus as to when and where libel, obscenity or hate speech occurs could be rather difficult, if not impossible, due to sharp cultural differences.

What should journalists and other online communicators make out of all this confusion? Perhaps two things: First, speech issues that have not exactly been easy to resolve in traditional media are profoundly more complicated on the Internet; they are also not likely to be neatly resolved anytime soon. Legal boundaries are likely to appear on the Net where none existed before, and those already in place may shift as competing interests—governments, the courts, media companies and civil libertarians—provoke and, ultimately, settle new disputes. Secondly, while it is important to sift through the legal complications, communicators should not be deterred from considering their own ethical responsibilities as users of the medium and members of the global community. If anything, the absence of clear legal boundaries in cyberspace should invite more dialogue about the values that underpin ethical communication and the best practices for applying them to the Internet. With that in mind, we'll look at several ethical speech dilemmas that have evolved as the Internet has expanded.

Ethical Dilemmas

Allowing Hateful, Harassing or Libelous Speech

The structure of the Internet, as it exists now, offers unique opportunities for journalists and other online communicators to interact directly with readers and customers. From the reader's perspective, the Net provides access not only to journalists but also to other citizens—the other members of the audience who are now more empowered to be active participants in what communications scholar James W. Carey calls the "conversation of our culture."[42]

On the Net, discussion forums, chat rooms, e-mail lists and newsgroups can serve as platforms for expression that are unequaled in scope by "letters to the editor" sections in newspapers or call-in segments on the radio. Many news organizations have parlayed their reader comment sections or Web "communities" into popular features that draw millions of visitors. But given the openness of the medium and the range of viewpoints held by participants, some of the expression advocates hatred or discrimination of people based on their race, nationality, religion, gender or sexual orientation.

Moderated or Unmoderated Forums?

What, if anything, should journalists, Web masters and their news organizations do about hate speech sent across their networks?

On the one hand, professional codes of ethics encourage journalists to use technology to foster public dialogue on controversial subjects. The Society of Professional Journalists, in fact, urges editors and reporters to "support the open exchange of views, even views they find repugnant."[43] It's likely that many journalists will find hate speech to be repugnant. If so, an argument could be made that they have an ethical responsibility to allow some measure of hateful speech to enter into the conversation taking place on their site.

On the other hand, many scholars and civil rights activists believe that hate speech, even if legal, is always unethical and should be condemned on moral grounds. Richard L. Johannesen, an ethics scholar and author of the book *Ethics in Human Communication*, argues that "racist, sexist, and homophobic communication . . . makes a person less than human." This type of expression, he asserts, "reflects a superior, exploitative, inhumane attitude of one person toward another, thus hindering equal opportunity for self-fulfillment."[44] On a similar note, author Haig Bosmajian contends that words—epithets, stereotypes, labels, and so forth—have historically been used "to justify the unjustifiable, to make palatable the unpalatable."[45] For example, slavery in the United States was often justified in language that labeled blacks as morally and intellectually inferior to whites. Likewise, the Nazis propagated, through the use of words and pictures, the idea that Jews were an inferior, even wicked, race that had to be eliminated—they used language to justify their unjustifiable crimes.

Would hosting an unmoderated forum on the Web lend an air of legitimacy to the values and beliefs that underlie hate speech? Would reporters, editors and Web masters be complicit in the spreading of unethical propaganda if they allowed racists, sexists or homophobes to post openly on their sites? Is it possible to promote an "open exchange" of ideas without contributing to the oppression of those targeted by hate speech? If so, what sort of protocol, if any, should govern the discourse?

Some journalists and civil libertarians suggest that unmoderated forums, which allow critics, and especially targets, to respond to hate speech, may have the effect of exposing the absurdity of such expression. From this view, more speech is far better than less speech, and what better medium to present all views than the Internet. This view is held by some "open publishing" sites and independent news portals.[46] Let the racists post, they say, and let the voices of justice, equality and peace drown them out. Besides, as an organization called Internet Freedom argues, "[f]ree speech puts a premium on the decision-making ability of each of us to weigh up all the arguments and draw our own conclusions."[47]

From a different perspective, some free speech advocates argue that some content that could easily be defined as hateful is socially and artistically useful. For example, social and political satire containing racial or sexual themes, and even epithets, may have the effect of shedding light on human hypocrisy or prejudice and encouraging informed dialogue on previously taboo subjects. One could easily cite Mark Twain's adventure novels or even the popular 1970s television show "All in the Family" as examples of pointed social commentary that have garnered praise as well as scorn. The Web enables a great many would-be satirists to present their writings and cartoons to a potentially wide audience. Should race, sex, nationality, gender and sexual orientation be off limits to humor?

Another view posits that news Web sites should not serve as a platform for hateful or harassing speech in any form. Allowing racist, sexist and homophobic comments, in the opinion of several major news organizations, would likely offend a large number of users; therefore, visitors are asked to adhere to a "code of conduct" before joining online forums. Moderators are also charged with removing offending posts. For example, the popular news site MSNBC.com "encourage[s] everyone to treat each other with mutual respect" and "reserves the right to remove [c]hat posts that advocate or encourage expressions of violence, bigotry, racism, hatred or profanity"[48] (see Figure 5.1).

The popular Web portal Yahoo! goes further, telling users that they should not "upload, post, email, transmit or otherwise make available any content that is unlawful, harmful, threatening, abusive, harassing, tortuous, defamatory, vulgar, obscene, libelous, invasive of another's privacy, hateful, or racially, ethnically or otherwise objectionable."[49] Earthlink, a popular ISP, takes a somewhat more libertarian approach, removing only hate speech that targets specific individuals.[50]

In the wake of terrorist attacks in the United States on September 11, 2001, many ISPs and news services reported that hate speech, especially against Jews and Muslims, increased dramatically. In response, they increased monitoring efforts and began to enforce speech guidelines much more aggressively. CNN, for example, doubled the number of moderators that screen incoming messages to its interactive sites. Yahoo! was particularly aggressive, removing a list of Arab-Americans, complete with contact information, which was posted under the heading "Islamic Terror Groups in the U.S." That action was quickly praised by many of the people who found themselves listed.[51] However, Yahoo! was also criticized for deleting messages from a student who criticized the U.S. bombing campaign against Afghanistan as a terrorist act. After repeated messages were blocked, the student complained that "[t]he Western media and politicians keep talking about their freedom of speech, freedom

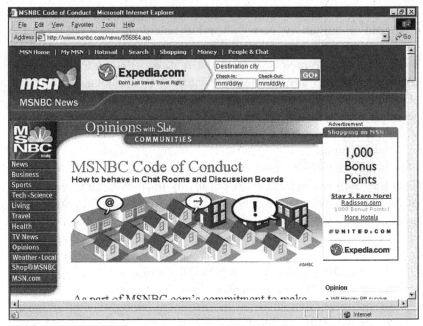

Figure 5.1 MSNBC.com's "Code of Conduct" for chat rooms and discussion boards.

of expression and how they are against the censoring of different views
. . . yet they are no different from any oppressive Third World country or
any dictatorship."[52] Yahoo! contends that it was simply reviewing posts
that generate complaints and removing them if they are deemed unlawful or harmful.

Perhaps the attitude of many Internet services and news organizations
after September 11th was summed up best by a spokesperson from Terra
Lycos, a popular Web portal and hosting service, "The sentiment in the
United States changed on September 11 about what's acceptable and
what's not in terms of what you can say."[53] That may be the case, but
some civil libertarians worry that self-censorship will go too far, causing criticism of government policies or institutions to be conflated with
speech that advocates violence against a nationality or ethnic group.
Additionally, there is the fear that those who edit the message boards
will impose their own biases on the discussion, filtering out views that
offend their cultural or political values, or caving in to pressure from advertisers or other powerful interests to remove content they find offensive, thus violating the SPJ code of ethics. Finally, free speech advocates

argue that, if anything, public debate, minus speech that clearly threatens violence on individuals, should be more open in times of war or national crisis. As historian Howard Zinn once wrote, "How ironic that freedom of speech should be allowed for small matters, but not for matters of life and death, war and peace."[54]

As with other ethical dilemmas, there are no hard and fast rules for deciding whether or not online forums and message boards should be moderated. Much will depend upon the values of the organization and the size of the staff—heavy editing will require more eyes to catch offending posts. It might be helpful, though, to consider several questions when adopting or altering a message board policy. For example:

- What are the possible consequences of allowing hateful messages —for users, for the site's reputation?
- What are the ethical reasons for allowing hate speech?
- What are the ethical reasons for banning hate speech?
- If forums will be moderated, what type of training and education— cultural and political—will the moderators be required to have? How will they decide what stays and what goes? How will hate speech be defined?
- How will posts actually be removed? Beforehand, based on a moderator's judgment, or after a sufficient number of complaints have been received? If the latter, how many complaints will be sufficient to warrant removal? Will the poster be informed immediately as to the reasons why his or her message was not displayed? How will repeat offenders be dealt with?
- Should anonymous messages be allowed? If not, will the site be guilty of cutting off a vital forum to people who need to keep their identities secret—that is, whistleblowers, crime victims, victims of spousal abuse?

With regards to libelous messages, it's important to remember that sites are responsible legally for content they moderate. If you moderate heavily, you bear more responsibility for what gets posted—you could face lawsuits. On the other hand, if you don't moderate, you face less legal risk, but you open up your site to the posting of offensive, hateful content and "flame wars" as well as informed opinion and commentary from users. Whether moderated or unmoderated forums are implemented, Nora Paul, a former instructor at the Poynter Institute for Media Studies currently serving as director of the Institute for New Media

Studies at the University of Minnesota, suggests that "the best protection is to make sure participants know what behavior is banned; guidelines should be consistently enforced."[55]

Linking to Offensive Content

Even if a Web site bans hate speech, pornography or other offensive content from its message boards, it is likely that journalists will find themselves reporting on such subjects. On the Web they will have the ability to embed links to background material into their stories. What, then, are the ethical implications of linking to commentary or images that readers might consider hateful or sexually offensive?

On the one hand, as the SPJ code of ethics indicates, journalists should not shy away from controversial subjects or "repugnant" views. If they are reporting on hate groups in traditional media, they might necessarily use quotations or sound bites from group members as part of their story. Providing a link to the Web site of a hate group might be considered equally necessary in online journalism. If readers are to know the full scope of racism, sexism or homophobia, shouldn't they be given easy access to important background material?

On the other hand, as we have noted, the Internet is quite different from print and broadcast media. Unlike quotations and sound bites that are carefully selected by a reporter from a larger supply of material before they are published or broadcast, hyperlinks can transport a reader directly to the full text of hate group propaganda. Likewise, links could bring users face to face with racist imagery or videos produced by hate groups. In a sense, links can send the reader into the virtual headquarters of the KKK or a neo-Nazi group. The case study in this chapter will attempt to shed additional light on this dilemma.

Linking to Grisly Images and Video

Also of concern is the ability to link to gruesome imagery related to major news stories. Print and broadcast outlets often refuse to show graphic scenes of murder or mutilation on the grounds that it would be offensive to viewers. On occasions when graphic videotape is shown on television news, it is usually preceded by a message warning viewers that they are about to see graphic pictures. If they wish to avoid such images or to protect their children from seeing them, they are urged to change the channel or turn the TV off. This is partly an acknowledgment of television's invasive nature, which was discussed earlier.

On the Web, though, journalists are able to link readers to graphic images or video without "invading" the private sphere—one must actively click the link in order to access the pictures. Based on this technological difference, online journalists affiliated with a print or broadcast operation could take advantage of the medium to publish or link to material deemed newsworthy, yet too offensive, for their traditional counterparts. It could also be easily promoted in print or on the air with a short message telling viewers or readers to "log on" if they wish to get more information and access controversial or offensive material.

Barb Palser, an online journalist and columnist for the *American Journalism Review*, wondered if TV stations and newspapers were using their Web sites in this fashion. She found that, to her surprise, many of them were not. According to Palser, online journalists have tended to treat graphic images, video, even profane language, in much the same way they treat them in traditional media. "Why are online news sites taking the high road when they could likely get away with so much more?" Palser asks, before answering her own question with, "Perhaps it's because they're beholden to the policies of their traditional parent companies, which have reputations to uphold and expectations to consider."[56]

Something of an exception occurred in June 2002 when *The Boston Phoenix*, a weekly newspaper in Boston, Mass., published a link on its Web site to a video that appeared to show the horrific murder of kidnapped *Wall Street Journal* reporter Daniel Pearl, who was killed and beheaded by radical Islamists in Pakistan in early 2002. In the video, Pearl comments on his Jewish heritage and appears to deliver a forced apology for U.S. policies in the Middle East. The three-minute clip ends with what looks like scenes of Pearl being beheaded by his captors. Before the *Phoenix* ran the link, CBS-TV aired an edited portion of the tape—which did not include murder scenes—on its May 14th evening news broadcast.

Steve Mindich, publisher of the *Phoenix,* defended the link on the grounds that the disturbing video "vividly illustrates the human toll of terror" and should be used to "galvanize every non-Jew hater in the world . . . against the perpetrators and supporters of those who committed this unspeakable murder."[57] Mindich also lashed out at the government for not making the video public months earlier when it was first discovered. A few days after the link was published, the *Phoenix* also published pictures taken from the video in their print edition.

A great deal of criticism followed, much of it expressing sympathy for the Pearl family, which had criticized CBS's earlier broadcast of the edited tape. In a statement released to the media at that time the family said, "We had hoped that no part of this tape would ever see the light of day."[58] Other critics argued that the *Phoenix* had only given Pearl's killers a wider platform for their twisted views.

Bob Steele, a member of the ethics faculty at the Poynter Institute for Media Studies, took a different approach, asserting that the *Phoenix* and its publisher had lost sight of the ethical principle "to avoid causing unnecessary harm" to those affected by tragedy. "Any journalistic purpose in publishing the photos of his death" so long after the event "is considerably outweighed by the emotional harm to Pearl's widow and family."[59]

Others suggested that the *Phoenix* was not necessarily wrong to publish the link, which was, after all, just a piece of hypertext that the user had to activate in order to enter the Web site that contained the video. In addition, the site clearly warned that the video was "extremely graphic" —users could choose whether or not to go forward. However, even critics in this camp suggested that, given the brutality shown in the video, the *Phoenix* could have placed additional features between the viewer and the video such as "a pop-up warning message with disclaimer that requires the user to select an OK button" before activating the link.[60]

Questions about the ethics of linking have also emerged in relation to news coverage of a controversial Web site at the root of a major court case. In 1997, anti-abortion activists created a Web site called The Nuremberg Files, to which they posted photographs believed to be of dead fetuses as well as "wanted posters" of numerous physicians who provided abortions, claiming that they were "war criminals." The posters also included home and work phone numbers and addresses of the doctors.[61] In light of past attacks on abortion clinics, Planned Parenthood sued the operators of the site, claiming that their content amounted to direct threats of violence against doctors and workers at abortion clinics. Adding to the threats, according to the plaintiffs, was the publisher's practice of marking out the names of doctors who had already been murdered or wounded by anti-abortion activists. In May 2002, a federal appeals court ruled that the wanted posters were "true threats" against the doctors and thus not protected by the First Amendment.[62]

In compliance with the court ruling the site operators removed the "strikethrough" markings from the wanted posters and stopped publishing addresses and phone numbers. They have continued, however, to publish graphic images reportedly showing dead fetuses. So the question arises, should online journalists covering this story link to the Nuremberg Files? If so, how should the link be constructed and what, if any, warnings or other information should accompany the link?

Case Study: Linking to Offensive Content

Hypertext links are perhaps the most fundamental feature of Web publishing; they allow writers to connect from one Web document to another or to connect from one location in the same document to another. Along with the servers that store hypertext documents and other multimedia files, links quite literally make the World Wide Web a living, global Web of interconnected information. And because HTML, the language used to create links, is accessible, flexible and easy to use, most any computer user can become a participant in the continuous building of the Web. Naturally, then, the medium offers opportunities for individual and collective expressions that were not present just a decade ago.

From a journalistic point of view, hyperlinks enable online reporters and editors to tap that vast reservoir of information and enhance stories by connecting readers instantly to source material, critical background data and diverse opinions. This new freedom is particularly useful when reporting on complicated or controversial political, economic and cultural issues that cannot be fully explored within the confines of traditional media. Before the advent of the Web, journalists could certainly mention the title and author of source documents (books, magazine articles and other media offerings), or they could print or broadcast the name, address and phone number of government agencies, businesses, community organizations or other news resources. But in order to access the information, the reader would have to switch from the TV, radio or newspaper to another medium (the telephone or the mail) or perhaps make a personal trip to the library or the source's location. Thus the reader would suffer an inconvenience for which the journalist had no remedy.

The Web and hypertext, of course, provide a remedy to this informational dilemma, but, like other liberating technologies, they also present some new dilemmas. Freedom on the Web means the freedom to post hateful, grisly or pornographic material as well as public records, scholarly work and informed political criticism and opinion. One link could expose readers to content they might want to avoid. Consider the following case in which a Web journalist researches and writes a story about racism and anti-Semitism.

The Case: One Click Away from a Hate Group

In February 2002, the Anti-Defamation League (ADL) issued a report accusing several white-supremacist groups of using the Internet to promote and distribute violently racist and anti-Semitic computer games. The report also indicated that hate groups viewed the slick computer games, which were aimed mostly at teenagers, as a major recruiting tool. Some of the games, featuring titles such as "Ethnic Cleansing" and "Shoot the Blacks," included scenes in which the lead character prowls an inner-city neighborhood looking for Jews and African-Americans to gun down.

The report quickly got the attention of Julia Scheeres, a reporter for Wired.com, the Web affiliate of *Wired* magazine. For the past year, Scheeres had been covering the online activities of hate groups and felt that the new report offered good material for another story. She contacted the ADL and interviewed the organization's national director along with an Internet researcher who worked on the report. Scheeres also interviewed the leaders of several white-supremacist groups and reviewed their Web sites. While building her story, Scheeres learned that many of the computer games mentioned in the ADL report could easily be purchased from the Web sites of hate groups. She also found that several sites offer free downloads of violent games in hopes that teens will access them and spread them around to their friends.

By the end of the day, Scheeres had completed her research and prepared her story for the Web. At 2:00 a.m. February 20, 2002, the finished product, titled "Games Elevate Hate to Next Level," was published on Wired.com. It included vivid descriptions of the computer games and numerous quotes from ADL representatives and hate group leaders. In addition, Scheeres and her editor decided to insert several hyperlinks to hate groups within the text of her article, effectively putting readers one mouse click away from racist and anti-Semitic material. Clicking one link in particular would connect users to a neo-Nazi Web site that offered free downloads of its "SS Man" computer game. The operator of the site is quoted in the article as saying that the games are like "free advertising" for his group, which aggressively recruits teenagers and young adults. No warning messages were placed next to any of the links.

Rationale and reaction. Scheeres says the decision to include the links was based on a simple protocol Wired.com follows for each story: "We always link to sites we discuss—no matter how controversial—unless they're illegal, such as child pornography."[63] Scheeres elaborates by explaining that Wired.com is a news organization and its job is to "inform the public," especially about issues relating to digital technology and Internet culture. Given the form of the medium, the liberal use of links is considered an essential feature of online reporting—it provides users with access to a wide range of uncensored information. If used this way, Scheeres believes, the Internet can have a liberating affect on both reporters and Web users: "We link to sites because we believe our readers are intelligent enough to make their own decisions about the material being discussed . . . [they] can chose to click on the hyperlink or not —we leave the decision up to them." Besides, says Scheeres, Web surfers could easily locate the sites using a good search engine.

Scheeres says reaction to her story was mixed—"some people didn't understand why we linked to the source material and others were outraged by them and didn't realize they exist." The day after the story appeared, one reader complained, in an e-mail to Wired.com, that the links may have helped the hate groups by giving them access to a much greater audience than they would normally have. As the reader put it: "traffic to your site is much larger than traffic to theirs . . . Great going guys." Scheeres also reports that some readers were so disgusted by the Web sites that they "threatened to hack them out of existence." Despite the negative feedback, Scheeres says she wouldn't do anything differently and adds that readers tend to complain more when Wired.com fails to provide links to Web sites and other background information.

Wired.com's approach to linking is rooted in the belief that legal information should flow freely through the Web. To filter out controversial or offensive words and pictures would be an act of censorship. However, several media critics and scholars argue that this type of reasoning is unsound because it equates ethicality with legality, tempting journalists to shorten or eliminate ethical reflection in favor of obedience to the law. Understanding legal boundaries is important, they argue, but consistently deferring to the law when confronted with professional dilemmas may be tantamount to abdicating social responsibility.[64]

From another angle, some journalists argue that Wired.com's approach incorrectly assumes that reporters and editors are primarily conduits for information who simply report the news without making value judgments. Aly Colón, a member of the Poynter Institute's ethics faculty, contends that the initial decision to report on any subject is an expression of a news organization's perspective and values. Similarly, the decision to insert a link marks another value judgment—it implies that the journalist believes the Web site being linked to is a source of information worthy of the audience's attention. Journalism, Colón adds, "involves making choices and taking responsibility for what is presented."[65]

When considering whether or not to link to a site, Colón advises journalists not to let the law or the medium make their decisions for them. Instead, he says they must first ask and answer the question, Is it valuable to users that I present this? If reporters and editors judge the site and its content to be newsworthy, they can move on to the question of whether or not the content is offensive, which presents a different challenge because tastes and sensitivities may vary from person to person. However, through consistent interaction with Web users, a good news organization can get a general sense of what its audience will find offensive. Letters to the editor, online forums, community dialogue and experience dealing with sensitive subjects can help editors gauge the sensibilities of their audience. With this knowledge in hand, online news editor Barb Palser says journalists can tackle the challenge of "deciding how to accommodate readers' sensitivities without limiting news content and resource functions."[66] In addition to knowing one's audience, Palser strongly advises journalists to consider the background of all sites they might link to. A few important questions should be in order, including: What other auxiliary content do they provide? Would users find that offensive? Who supports and finances the Web site? Are their publishing guidelines and user policies clearly labeled?

Linking options. It is quite possible that reporters and editors will judge a Web site to be both newsworthy and potentially offensive. When dealing with such a situation, ethics specialists remind journalists that the medium is flexible enough to accommodate a wide range of linking options—it doesn't have to be an all-or-nothing choice. For example, Web editors faced with an offensive Web site could easily do the following:

- Create an intermediate page between the article and the offensive Web site that explains the nature of the content and the news organization's rationale for inserting the buffer page.
- Include a brief parenthetical warning message next to a hot link. Users would still have easy access to the site, but they would be clearly warned in advance.
- Write the URL in the text of the story without providing a link. Users wouldn't be able to connect directly to the site, but they could find it later using the Web address.

If editors decide a Web site is far too offensive, they could also choose to do more storytelling—describing the material and the background of the group operating the site—rather than inserting a hot link or listing the URL. The ADL took this approach when it published its original report on the Web. Hot links embedded in the text directed readers to background material on hate groups and computer games prepared by ADL researchers.

Ultimately, solutions to the linking dilemma will vary according to each organization's journalistic and social values, the sensibilities of its audience, and the size of its editorial staff. Whatever the approach, consistency in applying it could serve journalists within the organization well; they would not have to navigate turbulent ethical waters without a guide. In addition, news organizations are advised to be transparent about their policies and to encourage vigorous discussion about journalistic practices—both internally with workers and externally with the user community. Explaining how and why ethical dilemmas are resolved could go a long way toward enhancing a news organization's credibility.

Policy Questions

Again, these are not easy questions to answer. Reporters and editors may have to balance their responsibility to give a thorough accounting of the story with consideration for the sensibilities of their readers. Perhaps many won't be offended by a link; they may not even wish to click it. However, others might be appalled if they don't feel they were given fair warning about what was behind the link.

Before making these judgments, online journalist might want to ask whether or not they have an ethical responsibility to protect their readers from content they might find offensive. Should they even attempt to determine what is or isn't offensive to readers? If they choose to link to questionable Web sites, should they at least warn readers about what they might see or hear? Reid Johnson, founder of a service that builds and maintains Web sites for television stations, suggests that even though readers must take affirmative steps to access most Web content, basic journalistic ethics dictate that warnings are called for. "I do think we have a responsibility to do a good job of advising the audience as they come into stories that there might be questionable or objectionable material found."[67]

As practical as written warnings might be, Steele cautions that they are not necessarily an ethical fix for the linking dilemma. "We cannot just say 'buyer beware,'" he told the *American Journalism Review*, "that alone does not mitigate against the harm that can come from tainted information." For Steele, simply issuing a disclaimer and washing one's hands after sending a reader to alarming content could be considered both "arrogant and irresponsible." The more ethical behavior, in his view, is for a news organization to make linking decisions based on its journalistic and social values and to take full responsibility for the consequences of such decisions.[68]

Posting Graphic or Offensive Images: Different Standards for the Web?

As noted, Web sites with affiliates in broadcast and print media are wrestling with the question of whether they should use a looser set of standards for publishing images and video on the Web. Based on an informal survey of Web sites in 2000, Palser reported that, for the most part, online journalists had carried existing standards to new media.

In the wake of September 11, 2001, the question was asked anew as photographers produced graphic images of the devastation of the World Trade Center. In addition to Palser, several other online journalists have suggested that graphic images not suitable for other media, such as the more graphic photos of September 11th, might be properly placed on the Web. J.D. Lasica, a senior editor with the *Online Journalism Review*, argues that different standards would be appropriate because

"Web users are extremely sophisticated about such matters," and as Palser notes, the medium is less intrusive than traditional media, broadcasting in particular.[69]

Before moving ahead with this dilemma, let's take a look at traditional standards for publishing photos in print. This is not exactly easy, as newspapers and magazines don't always share the same journalistic values—standards for judging what is or isn't appropriate will naturally vary throughout the news media. However, according to Phil Nesbitt, a former photography and design director for news operations in the United States and Asia, there are some general factors that a news outlet will base its decisions on: (1) the sensibilities of the community it serves, (2) the proximity of the event to its audience, (3) the direct relationship between the picture and telling the story (its news value), (4) individual bias, and (5) the public's right to know.[70]

These factors are much the same for broadcast news, though the time of day also becomes a consideration. For example, video that accompanies local reports dealing with sexual or violent themes is more likely to be broadcast in a late evening newscast rather than an early evening, midday or morning newscast. As one might imagine, coverage of live events complicates matters quite a bit. When events are being captured as they happen, journalists can't possibly anticipate every potentially offensive image that might be captured by cameras, thus they face the difficult task of making split-second decisions—whether to keep following the story and risk offending viewers or turn off the live feed and lose the story.

Until live video becomes commonplace on the Web, online journalists are more likely to face their dilemma from the perspective of those working in print media. The difference is that online journalists have a virtually unlimited number of pages in which to place images and the ability to build firewalls around sections with the most graphic photos.

This is what makes the prospect of altering standards for the Web so intriguing. Journalists might be able, it has been suggested, to present truthful, newsworthy images in a somewhat more tasteful way than simply tucking them on an inside page of the news section, where they might still cause alarm. In an e-mail discussion of this topic,[71] Steve Outing, a journalist who's worked for both print and online media, argued that publishing graphic images online doesn't thrust the content at readers; rather it allows them to make up their own mind as to whether or not they want to access it. "I'm not comfortable," Outing commented, "with deciding for people whether or not they should see such reality. What we can do on the Web is put them in a special area, with warnings that the images are graphic and some viewers may want to click elsewhere."

Outing specifically referred to a photograph of a severed hand, believed to be that of a September 11th victim at the World Trade Center, which most newspapers and Web sites refused to run (the *New York Daily News* was a notable exception, publishing the photo in one of its print editions). In a column written for the Poynter Institute's Web site, Outing argued that online journalists should have published the photo because it "demonstrate[s] the human toll" of the September 11th attacks in a way that photos of the planes crashing into the buildings or of body bags could not.[72]

Steve Yelvington, another journalist participating in the e-mail discussion, supported the idea of creating links from news items to special pages containing graphic images "so long as the nature of content is made clear by the context or title of the link to an item." Yelvington added that the primary reason for publishing any image should be "to inform and not to exploit."[73]

Other participants cautioned that publishing images such as the photograph of the severed hand could cause harm to family members of 9–11 victims, who number in the thousands. Even if they personally didn't access the photo, loved ones might be horrified to learn that the photograph of a body part, which might have belonged to a relative or spouse, had been posted on the Web where millions could see it, download it, and send it to others.

No doubt these arguments raise numerous questions: Is the cushion between online users and online content enough to warrant publishing graphic or offensive images? If so, how should pages containing the images be constructed—how thick should the cushion be? And what kind of warning should be posted at the entrance to the page? Does publishing questionable images carry with it more responsibility than linking to similar content on other Web sites? Finally, considering the SPJ's code of ethics, which states, "Journalists should . . . [a]void pandering to lurid curiosity," we should ask what would happen if, once posted, offensive images generate significant traffic. Will news organizations be tempted to publish more of them, even if they are of lesser news value? Should advertising be placed around the images?

As with other dilemmas facing online journalists, these may be difficult to resolve. As reporters, editors and their employers struggle to balance freedom with responsibility, they would do well to consider the following words from the book *Image Ethics*, which explores the moral issues surrounding the use of visual imagery:

> Photography, film, and television confer enormous power to create images that combine verisimilitude and visual impact . . . These powers are appropriately protected under our Constitution as an essential freedom in a democratic society, but they should also entail responsibilities. There is . . . the need for all concerned to pause and contemplate the moral implications of the images they produce and distribute.[74]

Complying with Governmental Demands

Pearl Video Redux

The online video of Daniel Pearl's murder brings up another ethical dilemma facing online journalists and Web publishers. How should they react to demands or requests from law enforcement or other government agencies to remove content?

On one hand, journalists often find it helpful to cultivate cordial, professional relationships with police officers and other public officials in order to get access to information and story leads. Additionally, journalists are citizens of their local communities and of their country; they are likely to sympathize with efforts of the police and other agencies to enforce laws and increase public safety. In this sense, quick cooperation may suit personal and professional needs.

On another level, however, journalists must be careful not to get so close and cooperative that they begin to behave as an investigative arm of government or compromise their First Amendment rights. Maintaining journalistic independence from those in power, both inside and outside of government, is, perhaps, the most difficult and important task each individual journalist and news organization faces. Without independence, journalists lose their ability to serve as public watchdogs.

In the case of the Pearl video, the FBI told a Virginia-based Web hosting company called Pro Hosters to delete the file from their server or risk being prosecuted under federal obscenity laws. News reports indicated that the FBI was acting on behalf of the Pearl family, which was asking Web sites to take down the video out of respect.[75] After receiving the FBI's request, the owner of Pro Hosters removed the video, only to repost it a week later after the American Civil Liberties Union (ACLU) advised him that it did not violate obscenity laws because the images did not appeal to the "prurient interest." The ACLU also consulted with the FBI and later reported that the agency's threats of legal action may have been a bluff. After reposting the video, the owner of Pro Hosters commented that his company "doesn't censor any type of content unless it's illegal." He also told *Wired News:* "I want it to be known that Pro Hosters is an advocate of freedom of speech and the press. We're not going to be strong-armed into not showing the video by the FBI."[76]

Legally, it appears that Pro Hosters was well within its rights to keep the video accessible from its server. But was their decision ethical? Should they have initially removed the video after the FBI threatened legal action? Moving beyond this case to a broader context, how should Web sites, especially those of news organizations, respond when asked by government agents to stop distributing content or to turn it over to law enforcement? For example, what if *The Boston Phoenix* had published the video on its Web site rather than simply linking to it? Would it be ethical for them to resist government demands to remove it? Should a news organization be held to a higher standard than a Web hosting company with no journalism connections?

This has long been a difficult dilemma for traditional media. As part of criminal investigations, court hearings or lawsuits, news organizations are often asked to hand material—reporters' notes, videotape, transcripts of broadcasts, and so forth—over to police. In particular, television newsrooms are often served with subpoenas to turn over their "raw" videotapes, though in some cases police have issued search warrants in order to obtain material they feel is needed to help solve a crime. Typically the matter is placed in the hands of lawyers. However, journalists may often consider whether or not ethical considerations should override legal advice. In other words, under what circumstances, if any, is it ethical to refuse to cooperate?

Defying a Court Order

Another question arises when Web publishers are ordered by a court to refrain from publishing content that has been deemed illegal.

The case of an online magazine called *2600*, which covers computer hacking issues, serves as an example.

In January 2000, a federal judge issued a temporary injunction against *2600* and its publisher, Eric Corley, which prohibited him from publishing an article containing the software code used to de-scramble DVDs so that they could be copied. Eight major Hollywood movie studios had sued Corley on the grounds that publishing the code violated the Digital Millennium Copyright Act (DMCA). As an act of "electronic civil disobedience" against the injunction and as a protest against the DMCA, Corley published links to Web sites that offered downloads of the software and urged others to spread the list of sites around. He contended that the First Amendment should protect his right to publish the links, which were merely addresses to other Web sites. The judge, who was not amused, made his earlier injunction permanent in August 2000 and further ordered Corley to stop linking to the sites,[77] which by this time had already been linked to by the *New York Times*, the *San Jose Mercury News* and several other news outlets.

In making the injunction permanent, the judge established a three-part test to determine when links from one Web site to another can be limited: (1) The publisher knows that illegal material is on the site being linked to, (2) the publisher knows that it is against the law to link to technology that circumvents the law, and (3) the publisher establishes the link "for the purpose of disseminating that technology." [78]

Several news organizations and civil liberties groups quickly came to Corley's defense, claiming that the judge's test could be used to undermine the ability of online journalists to connect readers to critical background information about controversial issues. Adam Clayton Powell III, director of the Freedom Forum, told the *Columbia Journalism Review* that "one of the most powerful informative elements" of the Web "could come under a cloud which could be construed as prior restraint."[79]

After two years of fighting the judge's ruling, during which time it was upheld by a federal appeals court, Corley decided not to take the case to the Supreme Court.

Aside from numerous First Amendment questions, the case also raises ethical questions; namely, how should one judge Corley's act of "electronic civil disobedience?" Was it ethical to publish links to software the court had deemed illegal? He certainly thought so, arguing that the de-scrambling software was a necessary fix for the DMCA, which he believed to be seriously flawed. If his actions were unethical, what should one make of the other news organizations that also linked to sites containing the software? Did the manner in which Corley attempted to circumvent the judge's order—openly calling on others to circulate the links—make his actions unethical?

It is worth noting the argument that not all legal behavior is ethical and not all illegal behavior is unethical. The United States has a long tradition of civil protest in which citizens have openly violated laws they believed to be immoral or unjust in order to draw attention to them. The American founders, in fact, were guilty of defying many English laws, including patent laws, in the run-up to the Revolution. Similarly, numerous civil rights advocates, anti-war activists and participants in other social movements have performed and defended illegal acts on moral grounds.

Whether or not Corley's actions are consistent with this tradition is certainly open to debate—one that can't be sufficiently addressed here. However, his case does provide an opportunity to highlight the belief held by many that one might disobey the law and be morally right in doing so. For online journalists who subscribe to this view, it would be helpful to carefully think through the circumstances under which they would consider resisting or defying the law to be ethically justifiable.

Summary

The right to speak and express oneself freely is considered by many to be the most sacrosanct of human rights. Free expression is a right protected by global treaties and is cited in international law.

In the United States, free speech—and its corollary the free press—is of course protected by the First Amendment. Many other countries have a similar constitutional protection, though some—such as certain European nations and South Africa, for example—specifically exclude racist and other hate speech from the broadest speech protections.

It's important to remember that the First Amendment only governs and restricts the actions of the government and not of private entities like corporations; this is the case because the founders of the United States were most concerned about checking abuse of powers by government and not by private companies, which, at the time, were not powerful economic and political institutions.

Over the years, there have been many attempts to bypass the First Amendment, one of the more famous attempts being the 1798 Sedition Act, which was ultimately ruled to be unconstitutional in 1964. There have also been several areas of speech that over time have been determined by the Court to lie outside of the First Amendments protections, called "unprotected speech." These unprotected areas are obscenity (a very specific legal term that is not the same as, say, indecency or pornography), incitement to violence, "fighting words," libel and defamation. Courts have also decided that certain restrictions placed on *how* free speech is exercised—most notably the "time, place or manner" of speech—are permissible, as are certain restrictions on commercial advertising.

The Internet surfaced several new First Amendment issues and has brought about new tests of freedom of speech in cyberspace. Congress has made several attempts to limit certain types of speech over the Internet. The CDA was an attempt to prevent the transmission of obscene and indecent messages or offensive content to minors. This bill was struck down as unconstitutional. A similar bill enacted in 1998, the COPA, is still being challenged and appealed as of this writing; the 2002 CPPA was struck down; and the CIPA is also still on appeal as of early 2003.

There have been other areas where the Internet has created a tension with the value of free speech. Notable areas are the Internet's capacity to facilitate political protest as well as the emergence of hate speech online. The Internet has also created hard-to-resolve disputes between countries on what is and is not legally permissible to communicate online, including clashes on hate speech and libel.

For online media professionals, the Internet has presented several specific speech-related ethical dilemmas. One is whether online media sites should permit offensive and hate speech on a public forum and how to develop appropriate policies and procedures to address such cases. Hate speech, for example, is not illegal, but is an unethical form of speech; so forum moderators may feel a responsibility to prevent insulting, denigrating speech. On the other hand, many journalists and others feel that suppressing speech—even repellent speech—is a bad solution. They believe that the best way to battle offensive speech is with more speech and are inclined to use their forum as a way to educate.

Another speech dilemma online regards the ethics of linking to hateful, violent, sexually explicit or otherwise offensive sites. Although journalists have an obligation to show readers the truth, they also have an obligation not to harm. A famous case that illustrated this dilemma occurred in June 2002 when *The Boston Phoenix* published a link on its

Web site to a video that appeared to show the beheading of kidnapped *Wall Street Journal* reporter Daniel Pearl. Some have suggested a solution to these kinds of cases whereby online news sites precede any linking to a particularly disturbing Web site with a warning. Similar dilemmas arise when an online news site considers posting gruesome or offensive images to its own Web site.

Finally, journalists are also forced to consider the ethical implications of whether and how to comply with demands from law enforcement or governmental agencies to remove content that has been determined by legal authorities to be illegal or forbidden.

Critical Thinking Questions

- You are running a forum on your site to discuss a recent article about affirmative action, and one of the readers begins using racist language. What principles are at stake? What is the ethical response to this situation?
- How is providing a link to an offensive site online different than a print newspaper publishing a disturbing picture or a television station broadcasting an offensive video? What are the implications of that difference in terms of the journalist's responsibilities?
- How should you determine whether a site on the Web is considered "offensive"? Is it better to draw up specific guidelines ahead of time or make the decision case by case? Explain.
- Does posting a warning to the reader that the site linked to is graphic or disturbing absolve the online media site of responsibility? Why or why not?
- Should online journalists consider the consequences of their readers going to the sites that they list? Should this be one of a journalist's ethical obligations? Why or why not?

Key Terms

First Amendment
Hate speech
Malice
Time, place, manner

Unprotected speech
  Fighting words
  Incitement to violence
  Libel/defamation
  Obscenity

Recommended Resources

Books

Meikeljohn, Alexander. *Free Speech and Its Relation to Self Government* (New York: Harper 1948), p. 27.

Mill, John Stuart. *On Liberty, Great Books of the Western World*, edited by R. Hutchins (Chicago: Encyclopedia Britannica, 1952), pp. 274–293.

Soley, Lawrence. *Censorship, Inc.: The Corporate Threat to Free Speech in the United States* (New York: Monthly Review Press, 2002).

Tedford, Thomas L., and Dale A. Herbeck. *Freedom of Speech in the United States,* 4th ed. (State College, PA: Strata, 2001).

Zinn, Howard. *Declarations of Independence: Cross Examining American Ideology* (New York: Harper Perennial, 1990).

Associations and Institutes

American Civil Liberties Union "Cyber-Liberties" page. Web page: www.aclu.org/Cyber-Liberties/Cyber-Libertiesmain.cfm.

American Library Association. Web page: www.ala.org/.

Electronic Frontier Foundation. Web page: www.eff.org.

Freedom Forum. Web page: www.freedomforum.org.

Index on Censorship. Web page: www.indexonline.org/.

Reporters Committee for the Freedom of the Press. Web page: www.rcfp.org.

United Nations, *Universal Declaration of Human Rights*. Web page: www.un.org/Overview/rights.html.

CHAPTER 6

Intellectual Property and Copyright

Chapter Goals:

- Provide historical background on copyright, intellectual property and copyright law.
- Explore the impact of the digital age on copyright.
- Identify ethical concerns of relevance to online media professionals: downloading copyrighted news content from the Web, linking to copyrighted material—"deep linking"—and creating an internal linking permissions policy.
- Examine the ramifications of noncompliance with existing copyright laws.

Does information really want to be free? What does this phrase really mean and how does it play out for online media professionals? This chapter examines the ethical dilemmas that grow out of issues surrounding copyright and intellectual property.

A popular conception of the slogan "information wants to be free" is that it implies that information shouldn't cost anything. Actually, the originator of the phrase, futurist and author Stewart Brand, used the word "free" to mean having "freedom," in that information should not necessarily be protected or secured. Furthermore, the phrase "information wants to be free" is only a portion of Brand's complete quote, which is reprinted here below:

On the one hand information wants to be expensive, because it's so valuable. The right information in the right place just changes your life. On the other hand, information wants to be free, because the cost of getting it out is getting lower and lower all the time. So you have these two fighting against each other.[1]

Brand expanded this remark in his book *The Media Lab*.

Information Wants To Be Free. Information also wants to be expensive. Information wants to be free because it has become so cheap to distribute, copy, and recombine—too cheap to meter. It wants to be expensive because it can be immeasurably valuable to the recipient. That tension will not go away. It leads to endless wrenching debate about price, copyright, "intellectual property," the moral rightness of casual distribution, because each round of new devices makes the tension worse, not better.[2]

Since the emergence of the Net, that slogan seemed to be ready made for use by that cadre of expert early Internet adaptors, commentators, futurists and writers then called by some the elite "digital digerati." The slogan was resurrected as a kind of "natural law" to justify and support a belief that the free information sharing that was occurring on the Net should be allowed to expand and flourish, without burdens of copyright, fees, restrictions, or other restraints.[3]

There is an intrinsic appeal to this point of view, in that much of what we call information, particularly the helpful knowledge-sharing type of information found on the Internet, seemed to simply be a type of fundamental human activity—like conversation, ideas and advice—that we do not normally expect to have to pay for and is set apart from the market. On the other hand, we are indeed accustomed to paying for certain types of information. There are instances where we feel that the creator of some information should be rewarded for putting in a great deal of effort, time and money; for example, articles, books, movies, research reports, music and other creative endeavors. Furthermore, most of us agree that it is even fair to pay for information transmitted orally—such as from a therapist or a consultant. Even some kinds of the most basic human information; oral conversation is accepted as information one pays for, such as going to a therapist or conversing with a consultant.

The most prominent clash between copyright, the Internet and ethics occurred when millions of people, mostly college students, began downloading copyrighted music files over the Internet, most notably via the now-defunct Napster service. Technically, all these people were probably breaking the law and engaging in copyright infringement, which would normally signify an unethical activity. However, there were some countervailing issues: There was some murkiness in the legal arena, the fact that so many freely and easily flouted a law called its appropriateness into question, and there were those who argued that such a law itself was unethical.

This chapter will provide a full discussion on this tension between copyright law and the Internet and look at it from the perspective of the online media professional.

Defining Copyright and Intellectual Property

The concept of *copyright* could best be described as a property right that recognizes the creator of an original published or recorded work (book, song, painting, etc.) as its owner, and as such, he or she has the exclusive right to control how that work is produced, reproduced (copied), and distributed. *Copyright laws* protect this right by defining what products are protected and which acts constitute infringement and by describing the mechanisms for enforcement as well as punishment for those who violate the law.

The term *intellectual property* is often used interchangeably with copyright, but it actually refers to several categories of property, or "tangible assets," that are created through the intellectual efforts of its creator or creators. This could include machines or other inventions protected by patent laws; trademarks, trade dress and trade secrets protected by trademark laws; and of course literature, paintings, music and other works protected under copyright laws. In other words, intellectual property refers to things while copyright refers to a specific kind of protection given to things that are published or recorded.

Before moving into a discussion of the ethical issues involving copyrighted works and other forms of intellectual property circulating through cyberspace, we should look back to the origins of copyright law and trace its evolution into the digital age.

Origins of Copyright Law

Although the beginnings of copyright law can be traced back to the emergence of the printing press, the concept of individual ownership of creative work may actually date back to ancient civilizations. The Greeks have been recognized for developing a distinctive sense of the individual, which unleashed cultural and intellectual talents, and may have helped spur the earliest claims of authorship. Scholars have also noted that the earliest cases of authors signing publishing contracts with manuscript sellers occurred in ancient Rome. Complaints about unauthorized recitation of poetry—plagiarism in the oral sense—may have also originated in Rome.[4]

The Church, the Court and Gutenberg

In medieval Europe, the Roman Catholic Church had a monopoly on the production of manuscripts, and hence, the communication of ideas, through its network of monasteries, which developed rules for the transcription and trading of manuscripts. However, monks did not claim authorship; that was reserved for the monastery, which was viewed as the province of God.[5]

Gutenberg's invention of the printing press in the 1400s made copying old documents vastly more efficient. It also gave enterprising printers the opportunity to publish new, original works, thus ending the Catholic Church's monopoly on the production and distribution of literature. Challenged by dissident religious thinkers, who could now disseminate their own views, the Catholic Church attempted to censor what it considered to be offensive texts.

Acting often in concert with the Catholic Church, European monarchs also moved to censor religious or political works that threatened their "divine rights." Assisting them in England would be the Stationers Company, a London-based printers guild, which was granted a monopoly on the book trade in exchange for helping the crown suppress seditious or blasphemous books and pamphlets. Anyone who wanted to publish legally had to obtain a license from the Stationers Company, which denied credentials to publishers who might offend the court. Reviewing this development, communications scholar Ron Bettig notes, "Copyright laws emerged simultaneously with censorship out of the trade practices of this monopoly."[6]

As literacy rates increased, readers demanded more and more original works of history, fiction and poetry in addition to the popular classical and medieval texts already in circulation. In response, publishers bought original manuscripts from contemporary writers or commissioned them to produce new works in exchange for money—the birth of the book contract. These transactions gave writers a sense that they owned their work and allowed them to earn a living outside the patronage system to which they had been previously bound. It also helped to break the Stationers Company's century-old monopoly on the book trade.[7]

Property Rights and the Statute of Anne

As writers gained a sense of individual ownership, political philosophers, most notably John Locke, reflected on the relationship between a person and the products created through his or her physical labor. Locke argued that working the soil of unused land should give one the right to claim that land as his or her property and to control how it would be used. Locke's philosophical descendants, including many of the American founders, developed a link between physical property and intellectual property—the products created by tilling the soil of one's mind—and subsequently argued that works of the mind deserved to be legally protected.

Following up on Locke's ideas about property and the utilitarian philosophy that authors might not produce new works unless they received some financial incentive, the English Parliament passed what is considered to be the first modern copyright law in 1710. The law, called the Statute of Anne, was labeled "An act for the encouragement of learning," owing also to the belief that a thriving publishing trade was essential to cultivating a vibrant, literate culture. The statute established a fixed copyright term of 21 years for existing works and granted a 14-year term to new works. At the time of expiration, copyright holders could file for a 14-year extension.[8]

It is important to note that the statute was narrowly focused. Rather than recognize the creative or moral rights of the author, it only prohibited the unauthorized copying of a registered work. Profits from a popular work were still likely to end up in the hands of the publisher—that much had not changed. However, knowing that whatever they published would be protected from piracy, authors were thought to be more likely to pay the author a better price *before* publication, when contracts were negotiated. Thus the law can be seen as an economic policy or trade regulation that benefited publishers primarily and authors secondarily.

Quite a different view about copyright law would develop nearly a century later in France and Germany. Philosophers in those countries argued that, indeed, authors had a moral right to claim ownership of their books, music and paintings. Supporting this justification for copyright protection was the belief that creative work is as much a product of the soul as the intellect and since both mind and soul are parts of the self, artists should be able to protect their creations as they would their own persons.[9] As dominant as this rationale would become in much of continental Europe, it did not take hold among American political architects, who were, not surprisingly, deeply influenced by the utilitarian economic views popular in England.

U.S. Copyright Law

In developing the U.S. Constitution, many of the delegates to the Constitutional Convention were convinced that copyright laws were necessary to promote the production of scientific and cultural works. At the same time, some delegates, particularly the leading architect of the Constitution, James Madison, were leery of state-protected publishing monopolies—or any economic monopolies for that matter—which they believed might stifle creative output. In hopes of striking a balance, the Constitution was carefully worded so that intellectual property rights would be recognized in a way that would serve the common good. Article 1, Section 8, Clause 8 states:

> Congress shall have power . . . to promote the progress of science and the useful arts, by securing for limited times to authors and inventors the exclusive right to their respective writings and discoveries.

With the phrase "science and the useful arts," the founders clearly intended copyright to protect technological, literary and artistic works that advanced learning and culture in the new country. And with the words "for limited times," they put forth the principle that copyright terms must not be so long that they keep important works from being used by the public. Above all, Madison and his colleagues believed that, as with the Statute of Anne, copyright laws must have as their primary goal the "encouragement of learning."

Empowered by the Constitution, Congress passed the first federal copyright law in 1790. It limited copyright terms for "maps, charts, and books" to 14 years with the right of renewal for another 14 years. It should be noted that the law prohibited only the copying of protected works produced in the United States, not the creation of new works, called "derivatives," based on copyrighted material or the copying of foreign works. Thus the first copyright law allowed Americans to make derivatives of domestic works and to pirate the works of foreign authors. As legal scholar Lawrence Lessig puts it, "We were born a pirate nation."[10]

The double standard was most likely Congress' way of trying to build something of a protective shield around the country's emerging cultural industry, while at the same time enabling American publishers to produce and disseminate cheap copies of more established European works, which could also be reworked or improved on by American writers and scientists. This is worth keeping in mind when considering the current efforts of people in developing countries to make and distribute inexpensive derivatives of American goods such as agricultural products, medicines and communications technology.

Copyright Expands

Throughout the 19th century and the first half of the 20th century, American copyright law expanded to include new media formats such as music (1831), dramatic compositions (1856), photographs (1865), works of art (1870), recorded music (1909), motion pictures (1912), and nondramatic literary works (1953). Copyright terms were also lengthened during this time period, from 14 to 28 years (1831) with a renewal term of 14 years, which was then extended to 28 years (1909).[11] In addition, copyright holders were given the right to control adaptations, translations, or other derivative works (1870), and the law was updated to establish copyright relations with foreign countries (1891).

Industry leaders who felt that new technologies would facilitate the copying, or "stealing," of their works vigorously sought most of these expansions. Their rationale was that existing law didn't adequately address the threats presented by new media; therefore, copyright protection should be extended to cover new formats, and terms should be lengthened so that owners would have more incentive to produce future works.

Jessica Litman, a law professor and author of the book *Digital Copyright*, contends that under pressure from industry leaders, Congress has tended to pass new, increasingly complicated laws that give copyright holders greater and greater control over media content, even going so far as to let them participate in negotiating sessions where new rules were hammered out. At the same time, consumers and their interests have not been represented at the bargaining table.[12] Ultimately, according to

Lessig, this has moved the law further and further away from its constitutional, and perhaps ethical, purpose—that of stimulating scientific and cultural production for the "encouragement of learning" and innovation. Instead, he argues, much of our once-common culture has been turned into chunks of private property ("propertized" as he calls it) and locked away to the detriment of current and future artists, authors and scholars.[13] Litman further observes that copyright owners are not usually artists and authors themselves but large media companies; therefore, the law protects industry far more than it protects the actual creators of scientific and cultural goods.

Copyright Act of 1976

More recently, copyright law went through a major congressional overhaul in 1976, which resulted in passage of the act that forms the backbone of copyright law today. This continued the trend of broadening the rights of copyright holders while including numerous narrowly defined exceptions, designed primarily to accommodate interests such as movie theaters, TV/radio broadcasters, educators and libraries that use copyrighted works. In a sense, then, copyright law had turned 180 degrees since 1790 when the rights of copyright holders were narrowly defined and the rights of others to adapt, translate and share scientific and cultural products were quite broad.

Among the specific issues addressed by the sprawling 1976 act are definition of what qualifies as a copyrightable work; terms of copyright ownership and transfer; duration of copyright, which was lengthened to the life of the author plus 50 years; copyright registration procedures; definitions of copyright infringement and remedies for redress; and duties of the U.S. Copyright Office.

As for what works can be copyrighted, the act states that only "original works of authorship" that have been published in a "tangible" form of expression are eligible. Thus it is the expression of an idea, rather than the idea itself, that can be copyrighted. In using the word "tangible," the act covered all contemporary media formats and tried to anticipate formats that might emerge in the future. This includes literary works; musical compositions, including lyrics; dramatic works, including accompanying music; pantomimes and choreographic works; pictorial, graphic, and sculptural works; motion pictures and other audiovisual works; sound recordings; and architectural works.[14] The act also states that unpublished works are covered. For example, your personal letters and diaries become your property the moment the ink hits the page or the moment you type words on paper. Word processing documents count as well; they become fixed when they are printed or stored on a disk or on your hard drive.

Under the act, copyright holders have the exclusive right:

- *To reproduce* the work in copies or phonorecords;
- To prepare *derivative works* based upon the copyrighted work;
- *To distribute copies or phonorecords* of the work to the public by sale or other transfer of ownership, or by rental, lease, or lending;
- *To perform the work publicly*, in the case of literary, musical, dramatic, and choreographic works, pantomimes, and motion pictures and other audiovisual works;
- *To display the copyrighted work publicly*, in the case of literary, musical, dramatic, and choreographic works, pantomimes, and pictorial, graphic, or sculptural works, including the individual images of a motion picture or other audiovisual work; and
- In the case of *sound recordings, to perform the work publicly* by means of a *digital audio transmission.*[15]

The 1976 act is also careful to spell out what cannot be copyrighted. This includes artistic ideas not yet fixed in a tangible form of expression; titles, names, short phrases and slogans (though these can be registered and protected as trademarks); facts, ideas, procedures, methods, systems, processes, concepts, principles, discoveries or devices, as distinguished from a description, explanation or illustration of them; and works consisting entirely of information that is common property and containing no original authorship (for example, standard calendars, height and weight charts, tape measures and rulers, and lists or tables taken from public documents or other common sources).[16]

Fair Use

Perhaps most importantly for the public, the 1976 act also enshrined the "fair use" doctrine. Simply put, this allows users to make and distribute copies of protected works provided they are for "criticism, comment, news reporting, teaching (including multiple copies for classroom use), scholarship, or research."[17] This is the principle that allows movie reviewers to broadcast a scene from a film. It also allows us to quote from other authors in this book!

The concept of fair use as a limitation on copyright and a defense against charges of infringement can actually be traced back to a court case in 1841. On the question of whether or not the use of George Washington's private letters without permission from the owner of the letters constituted copyright infringement, the court ruled that it did not. Their rationale was that under certain circumstances, copyrighted works could be used in a manner that did not harm the owner. The key, as a judge wrote in the court's decision was the "nature and objects of the [copyrighted] selections made, the quantity and value of the materials used, and the degree in which the use may prejudice the sale or diminish the profits, or supersede the objects, of the original work."[18]

The current fair use test is remarkably similar to the one laid out by the court in 1841. As the Copyright Act of 1976 states, the following four factors must be taken into account:

1. the purpose and character of the use, including whether such use is of a commercial nature or is for nonprofit educational purposes;
2. the nature of the copyrighted work;
3. the amount and substantiality of the portion used in relation to the copyrighted work as a whole; and
4. the effect of the use upon the potential market for or value of the copyrighted work.[19]

Unpublished works are also subject to the fair use exception "if such finding is made upon consideration of all the above factors."[20]

Fair Use and Journalists

The fair use doctrine is extremely important for journalists who, like teachers and researchers, must be able to use parts of copyrighted works in order to do their jobs effectively. Using the four fair use factors as a guide, media attorneys typically advise reporters and editors that they must take care to use only portions of copyrighted works that are the subject of a news story, that they should use no more of the material than is necessary, and that they cannot take the commercial heart of the work. To borrow an example used by Lynn Oberlander,[21] a media attorney and college professor: A news report about the assassination of president John F. Kennedy that refers to the Zapruder film may use a very short segment of the film (perhaps a few seconds' worth), but only so long as the segment doesn't show the president being shot, which is considered the commercial heart of the film—that's where its value comes from. The best approach would be to use a few seconds from the beginning of the film.

When dealing with printed works, similar rules apply. Reporters can quote from copyrighted works so long as they stick to short sections that specifically relate to their news article or commentary and that could not be considered as the most commercially viable segments of the work. Quoting from a passage in a murder mystery in which the author reveals the identity of the killer would not be a good idea.

As one might imagine, it is not easy in many cases to make a determination about how much content is fair use. Investigative reporters who obtain unpublished material such as private memos are especially susceptible to being sued for copyright infringement by sources or subjects who may seek an injunction to prevent the story from being published at all. Though unpublished works are included in the fair use doctrine, courts have been reluctant to give reporters as much leeway to use them without the owner's permission.

Parodies

The courts have also ruled that a parody of a copyrighted work in which some portion of the work is used may qualify as a fair use. The most important legal case in this area involves a musical parody of the Roy Orbison song "Oh, Pretty Woman" recorded by the rap group 2 Live Crew. The Supreme Court ruled that even though key riffs from the original had been used and the parody was made for commercial purposes, it did qualify as a fair use because the amount of music copied was necessary to "conjure up" the original work and the market for the parody was not the same as the market for the original.[22]

First Sale

One other copyright limitation worth noting is the "first sale" doctrine. This gives consumers the legal right to sell, rent or lease a copyrighted work to whomever they choose. In other words, once you've purchased a book, you can read it as many times as you like, or you can loan or sell it to a friend. Making copies of the work would still constitute infringement, but barring that, you can do pretty much whatever you want after you've purchased a work. This is in keeping with the principle that copyright law should only regulate the production and distribution of protected works, not the *consumption* of them. As we'll see below, some critics argue that digital copyright laws may disable this important exception.

Public Domain

Before moving on to digital copyright protection laws, it is important to mention that when all copyright terms expire, a work passes into the public domain where it can be freely copied, translated, adapted and distributed, except in cases where the work is still protected as a trademark —for example, Beatrix Potter's *Peter Rabbit* illustrations. The public domain also includes all U.S. government works, state judicial opinions, and legislative enactments and official documents as well as all of the other things listed previously that are not copyrightable.

The public domain, in effect, can be viewed as the place where cultural and scientific works flow freely; for example, the place where a musician might locate an old folk or bluegrass tune, rearrange the chords a bit, rephrase the lyrics and present it anew as a transformative work. One only has to consider the success of the soundtrack to the film "Oh Brother, Where Art Thou?" to realize what a valuable reservoir of culture the public domain is. Contemporary bluegrass pickers can delight audiences with new recordings and live performances of a song such as "Man of Constant Sorrow" only because the song is freely accessible to them. However, many critics of copyright law argue that Congress' tendency to keep lengthening copyright terms—for works already created as well as those not yet created—keeps vast amounts of culture from entering the public domain, effectively preventing current and future generations from doing what many of the artists on the "Oh Brother" soundtrack were able to do.

Copyright in the Digital Age

The Copyright Act of 1976 was amended several times during the 1980s and '90s to protect new media works and to bring the United

States into full compliance with the Berne Convention, an international treaty that governs copyright in some two dozen major content-producing countries. One of its provisions eliminates the requirement of a copyright notice on a copyrighted work. This, in effect, means that a copyrighted work is protected even if the owner does not display the (c) copyright symbol on it. Most intellectual property attorneys still advise clients to display the symbol on copyrighted works as a precaution in case of possible infringement. The symbol's presence can be used in court to prove that the defendant had knowledge that the work was protected before he or she used it without permission.

This brings up another important point about copyright protection and attempts to sue for infringement. Just because all published and unpublished works are considered to be the property of their creators the moment they are fixed in a tangible form doesn't mean that the owner can sue successfully for infringement. The work still has to be registered with the U.S. Copyright Office in order for the owner to sue for and receive damages. Registration is a relatively easy process that can be completed at any time during the copyright term. The key factors in getting a registration approved are the originality and creativity of the work. It's not enough to discover and present some facts; the author or artist must interpret or arrange those facts in a creative and original manner. An entire Web site can even be registered if it passes this test. Web designers should be aware that creating an unauthorized adaptation of another site could be construed as infringement.

Digital vs. Analog

Digital copyright extensions have been driven by more than a desire to comply with international treaties. Major copyright holders such as movie studios, record labels and software manufacturers expressed fears that the emergence of the Internet and the World Wide Web would allow users to pirate digital content much easier than works fixed in traditional media formats. Additionally, they feared that traditional works could be scanned or converted into digital files and distributed online—old formats newly digitized could converge on computer-based platforms to the detriment of copyright holders.

In this sense they saw the Internet as a giant copying machine whose decentralized architecture would enable millions to efficiently trade digital content on a global scale. As Jack Valenti, the chairman and

CEO of the Motion Picture Association of America (MPAA) put it, "digital is to analog as lightning is to the lightning-bug." Valenti also considers file sharing to be nothing more than "digital thievery."[23] Without additional copyright protection, Valenti and others have argued, content owners would suffer financially, thus losing their ability to fund new creative works.

In one of the first responses to their lobbying pressure, Congress amended the Copyright Act in 1990 to prohibit the commercial lending of computer software, which significantly modified the first sale doctrine. No longer could a user legally sell, rent or lease software that he or she has already purchased. There were some exceptions for libraries, which may loan software provided they place a copyright warning on the programs. Software makers who supported the revision argued that it was necessary to curb the pirating of software. They estimated that as much as 50 percent of all copies of their products used in the United States were unauthorized.

Two years later the act was amended again to make copyright renewal automatic, whether or not the owner intended to keep selling the work. This significantly curtailed the flow of works produced before 1978 into the public domain.

Random Access Copying?

In perhaps their boldest and most ingenious move, attorneys representing major copyright holders argued that anytime a user loads a computer program or views a file online they have made a reproduction, or copy, of that work in violation of copyright law. As law professor Litman explains it, this argument was based on the fact that a computer reproduces things "in its volatile Random Access Memory [RAM], and anything that exists in volatile memory could, at least in theory, be saved to disk." This being the case, the attorneys argued that copyright owners had the legal right to "collect money for every single appearance of a work in the memory of a computer anywhere," according to Litman.[24] If Congress and the courts accepted such an argument, copyright law could then be used to regulate the *consumption* of legally purchased content in addition to the production and distribution of it.

As far-fetched as this argument might seem, it worked on several levels. First, the courts agreed with the basic premise and applied it to a 1993 case, ruling that unauthorized downloading of software to RAM was the same as making a copy in a "fixed" form, which violated the Copyright Act.[25] This reversed a 1984 decision, in which a court held that data stored in RAM was not a copy as defined by the Copyright Act.[26] Secondly, as Litman reports, the argument alarmed ISPs and other stakeholders who often stored and transferred parts of copyrighted works on their networks, convincing them to support the rationale if they could be granted immunity from liability in any new copyright laws.[27] Finally, the major copyright holders were able to convince Congress to amend the Copyright Act to provide what they saw as adequate protection for their works in the new media environment.

Criminalization, Copy Protection Technology and Term Extensions

In the late 1990s Congress passed three major laws that expanded copyright protection. The first of these, called the No Electronic Theft (NET) adjusted the penalties for copyright infringement, making some offenses a criminal rather than civil matter. For the first time, copyright violators could go to jail, perhaps for as much as 5 to 10 years, instead of being forced to stop making copies and pay a fine.

The Sonny Bono Term Extension Act (CTEA), passed in 1998, extended the term of copyright protection to the life of the author plus 70 years, which is consistent with the European Union's terms. Works for hire are protected for 95 years, up from 75 in previous versions of the Copyright Act. These extensions were applied retrospectively, further extending the terms of works already created.

This CTEA was challenged in court by a New Hampshire man, Eric Eldred, who publishes great works of literature—including works by Nathaniel Hawthorne, Joseph Conrad, Henry James and Louisa May Alcott—on his Web site (www.eldritchpress.org), which has been recognized by the National Endowment for the Humanities as an exceptional Web resource. Eldred had planned to publish several works that were set to pass into the public domain in 1999, but his plans were derailed by the CTEA, which extended the copyright on those works until 2019. In response, Eldred, with the help of several prominent attorneys including Lawrence Lessig, filed a lawsuit claiming that the act was unconstitutional because it extended copyright terms well beyond the "limited times" that the Constitution called for. In a 7 to 2 decision the Supreme Court ruled against Eldred and his fellow petitioners, arguing that Congress did not violate the Constitution's copyright clause when it extended copyright terms under the CTEA.[28] In an interesting twist, 17 noted economists, including several Nobel Prize winners, had filed a brief on behalf of Eldred arguing that the costs to the public—increased prices for books, songs, movies, and so forth—brought on by the copyright extensions are far greater than any financial benefits that would accrue to copyright holders.[29]

Digital Millennium Copyright Act

The third measure, known as the Digital Millennium Copyright Act (DMCA), has probably been the most controversial. Following the RAM rationale described previously, it amends the Copyright Act of 1976 to prohibit the manufacture, sale or use of any program designed to circumvent technologies that protect copyrighted works. Thus if a company produces digital versatile discs (DVDs) that are encrypted (scrambled) so that they cannot be played on certain computer platforms, it is illegal for a consumer to use a decryption (de-scrambling) program to make one of these discs playable on his or her computer. At least one person in the United States has been arrested and indicted for selling circumvention programs.[30]

The provision against circumvention applies even to cases in which the use of the content would qualify as a fair use, such as a teacher decrypting a DVD or other encrypted item in order to use it in class for educational purposes. The DMCA, in fact, contains no explicit exemption for fair use. It does, however, have some narrow provisions that allow law enforcement agencies and software companies to gain access to protected works in order to investigate crimes and perform computer maintenance.

The courts have ruled that under the DMCA it is also illegal to distribute a decryption program to others by posting it on the Internet or by posting a link to a Web site that contains it (this is the *2600* case mentioned in chapter 5). For many critics the act's provisions and court interpretations effectively disable both the fair use and first sale doctrines for digital media products— in their opinion a serious flaw that constrains the ability of consumers, educators and the public at large to use content in ways that are legal offline.

The DMCA also creates a "safe harbor" for ISPs, which allows them to avoid liability for copyright infringement under certain conditions. This is part of the deal struck between copyright holders and network operators that professor Litman discusses in *Digital Copyright*. The provision was also guided by the rationale that ISPs are primarily conduits for millions of packets of information that are transmitted to and from users. It would be virtually impossible for ISPs to monitor all messages traveling through their networks; therefore, they should not be held liable for copyright infringement they could not reasonably prevent. Supporters of the act argue that by protecting many Internet companies this provision offsets other weaknesses, making the DMCA, on balance, "a useful law that benefits e-commerce" rather than "just a sword the government can use to slice away at fair use."[31]

Operators of Web sites and message boards are not covered by the safe harbor provision, though, and may be held liable for "contributory infringement" if they know a user is copying and transferring files and they fail to take appropriate action or they are careless in monitoring their site for copyright violations.[32] This is what got Napster, which was actually established to facilitate the transferring of music files, in trouble and led to its demise.

Additional Proposals and Digital Copyright Issues

There are several other issues regarding digital copyright that are worth covering briefly before moving on to specific ethical dilemmas faced by journalists and other online communicators using copyrighted material.

The Consumer Broadband and Digital Television Protection Act

The Consumer Broadband and Digital Television Protection Act (CBDPTA), proposed in 2002 by Sen. Ernest Hollings, would require consumer electronics manufacturers to embed government-approved security (copy protection) into all digital devices including personal computers, digital cameras, cell phones, fax machines, MP3 players, PDAs and DVD and compact disc players. Supporters from major movie studios and record labels argue that the act is needed to prevent users from pirating software programs and entertainment content. Opponents, including some electronics manufacturers and numerous consumer groups, contend that the act would turn technology into digital police and be extremely costly to implement in addition to preventing consumers from using digital media players the same way they've used analog VCRs and audiotape recorders.[33]

Database Protection

Traditionally, the compiling of data or other publicly accessible facts, such as phone numbers or addresses, into a database has not been considered copyrightable except in cases where the data was selected and arranged with originality and creativity.[34] However, there have been attempts to steer around this tradition by proposing laws that would, according to legal scholar Yochai Benkler, create "a property right in compiled raw data."[35] In 1996, the European Union passed a law that gives database owners copyright protection. If Congress were to pass such a law, Benkler asserts that it would raise the cost of data to a point where many educators, amateur producers and nonprofit organizations could no longer afford access to it.

Web Radio Fees

The DMCA required online radio stations to pay royalty fees (of 7 cents per song, per listener) for copyrighted music played on their Webcasts. But most Web radio stations, which tend to be operated by colleges, universities, religious organizations, and small, independent broadcasters, have tiny budgets and could not afford the fees. Many were faced with the prospect of shutting down their operations. Several Webcasters, in fact, did go silent in 2002 as the fees were about to be implemented. By late 2002, however, Congress had passed the Small Webcasters Settlement Act (SWSA), which suspended all royalty payments for noncommercial stations for six months and invited Webcasters and the recording industry to negotiate a new royalty system that is more acceptable to both sides. National Public Radio (NPR) had already negotiated its own undisclosed deal with the industry prior to passage of the SWSA.[36]

Ethical Dilemmas

In light of the recent, and possibly future, revisions to copyright law in the digital age, some critics have argued that copyright protection "no longer has a consistent theory, let alone an ethical position."[37] Others, such as professors Lessig, Benkler and Litman, have suggested that the law no longer attempts to strike a balance between protecting the rights of creators and the rights of the public to have access to cultural and scientific works, making the law unconstitutional and, perhaps, immoral.

Lessig and Benkler, in particular, argue that digital copyright protections are so lopsided in favor of content-owning media conglomerates trying to get "perfect control" over their products that they threaten to undo the potential for "peer-to-peer" sharing and innovation the original architecture of the Internet made possible. If not reversed and brought back to a sensible balance, they contend, copyright law will only allow the media powers of the previous era to veto the creative promise of the future.[38]

From the opposite side, leaders in the entertainment industry argue that copyright extensions uphold ethical values of fairness and just compensation by providing protection for creators who must now contend with greater copying threats than ever before. As mentioned above, industry leaders consider file sharing to be thievery deserving of criminal punishment. In testimony before Congress, Jack Valenti said the U.S. movie industry "suffers revenue losses of more than three billion dollars annually through the theft of videocassettes." Unless stringent measures to protect digital content, which can be pirated much easier than analog products, are in place and backed up by law enforcement, Valenti contends that "the entire fabric of costly creative works is in deep trouble."[39]

These contending views will percolate through most, if not all, of the ethical dilemmas described in this section. In addition, the arguments of other scholars and technology experts will be introduced as the discussion moves into dilemmas specific to their interests and expertise. Readers will note several situations that have particular relevance to journalists while others pertain to those involved in Web design or Internet users in general. As with laws discussed in previous chapters, there are some complicated and conflicting aspects to copyright law, which make ethical reasoning and decision making even more important. And as we've just noted, some critics contend that digital copyright laws may even be unethical, and they've asked, as we will too, whether or not active noncompliance is an ethical position.

Downloading/Uploading Copyrighted Material

Perhaps the trickiest part about dealing with intellectual property issues in cyberspace is the nature of the property in question—that is, the form of digital music, movies, computer games, software, even Web pages. Though they are considered to be fixed in a tangible form, they appear to be so much more intangible, or virtual, as opposed to physical. A computer program, for example, is basically a sequence of instructions—a recipe, if you will, that tells hardware and other software what to do. Further, digital data can be stored on disks, hard drives or Web servers. They can even be broken into tiny packets of data and whisked around cyberspace. Thus it seems hard for many people to accept the claim that clicking a link or browser button in order to download a stream of code is the same thing as stealing physical property. As an article in *The New York Times* put it, "For all the work that goes into designing these digital creations, they just don't quite seem like real things."[40]

Piracy or Ethical Sharing?

Given the nature of the content and the architecture of the medium, can the act of downloading copyrighted content without paying for it, or without getting permission from the owner, be construed as "thievery" or "piracy" as the record companies, movie studios and major software manufacturers call it? Does such an act bring harm to the artist or author? If so, downloading a protected work could be viewed as both an illegal and unethical act. It would also appear that millions of Internet users are breaking the law and behaving unethically everyday. Most of them may be downloading the material for personal use only and don't plan on distributing it to anyone else for commercial gain. Should this make any difference as to whether or not the behavior is illegal or unethical?

Let's now flip the issue around a bit. If one is downloading files from the Internet, someone else must have uploaded them. Is there a fundamental difference between the two acts? Is one more or less ethical than the other? Many Web users argue that uploading content that one has already purchased is the online equivalent of sharing, renting or selling a book, video or CD with someone offline, thus it falls under the first sale doctrine and should be considered both legal and ethical. However, there are some significant differences between physically loaning a CD to one's friends (or even making a copy and loaning that to one's friends) and putting it on the Internet where it can be downloaded by every user. This is the primary concern for copyright holders: That one upload could have the effect of giving free copies to millions, thus discouraging them from buying their own copy. Also, as we've noted, the DMCA does not include an explicit fair use or first sale exception for consumers, and in the Napster case, a court ruled that people using the service to share files, even for noncommercial purposes, were violating copyright law. Napster was also found to have committed contributory infringement because its service was used for illegal purposes with the knowledge of its operators. Should the legal ruling and/or the reach of the Internet affect your views about the ethics of file sharing?

Alternative Approaches: Three Viewpoints

On another level, we might even ask whether traditional concepts of intellectual property and copyright are even applicable to the Internet. Should the law, as legal scholar Litman suggests, move away from treating the reproduction (sharing or copying) of a work as copyright infringement and, instead, focus on stopping the unauthorized *commercial exploitation* of protected works? If so, then one might argue that sharing files online is ethical and should be legal, so long as the parties involved aren't distributing hundreds or thousands of unauthorized copies for commercial gain.

Ethics scholar Edwin C. Hettinger takes a more radical approach, questioning the common justifications for intellectual property in the first place. He contends that unlike physical objects, intellectual products can be used by more than one person at the same time and do not typically dissipate with use. He also views copyright laws as being contradictory to political values such as freedom of expression. If a society values free speech, how can it also create laws that treat the expression of ideas as private property? On the argument that intellectual labor must be rewarded, Hettinger counters that it is difficult to assign property rights to one individual creator because intellectual products are built on the labor of others. In other words, creative work doesn't exist in a vacuum; it is influenced by countless sources. Finally, Hettinger refutes the claim that copyright laws encourage the production of new works. Instead, he suggests that they slow down the dissemination of knowledge, leading him to wonder "why one person should have the exclusive right to possess and use something which all people could possess and use concurrently?"[41]

Using this rationale, online services such as Napster would be ethical because they contribute to the efficient diffusion of intellectual and artistic works, which will influence and inspire others to create more products, further adding to the cultural stock and serving the greater social good.

On a somewhat similar track, noted software developer Richard Stallman, a driving force behind the "free software" movement, argues that digital copyright laws do more harm than good because they restrict ethical uses of digital content by consumers.[42] He also takes issue with use of the term *piracy* to refer to the act of file sharing. "To describe a forbidden copy as pirate is a perversion of morality," he argues. "It's designed to convince everyone that sharing with your neighbor is the moral equivalent of attacking a ship." Following this logic, he fully supports noncommercial file-sharing services, which he says are "so useful that [they] must be permitted: you can't tolerate giving up the freedom to do something so useful."[43]

Still, Stallman does see the need for some type of basic copyright protection, only his model would treat works differently based upon their uses. For example, he identifies three broad categories of intellectual property: (1) *functional works* used to accomplish a task; (2) *representative works,* which are an expression of someone's intellectual ideas or opinions; and (3) *aesthetic or entertaining works,* which are designed to create a particular sensation. Stallman puts computer software in the first category and suggests that there should be little or no copyright protection for them—perhaps three- or four-year terms. For works in the second category, he recommends allowing them to be copied only verbatim—so as not to misrepresent the views expressed—for noncommercial purposes. And for the last category, Stallman suggests copyright terms of 10 to 20 years and a system whereby those who wish to make commercial adaptations would pay fees to the author or artist.

Circumventing Copy Protection

In addition to lobbying for broad new copyright laws, the entertainment and software industries have put a great deal of effort into developing technology designed to proactively stop users from copying or playing their works in certain machines. One popular technique is to encrypt, or scramble, DVDs so that they can only be played in DVD players or computers running only Macintosh and Windows operating systems. As a result, DVDs encrypted with the Content Scrambling System (CSS) method will not play in computers running the Linux oper-

ating system, thus Linux users are prevented from playing legally purchased DVDs in their computers.

In response to CSS, a group of hackers developed a de-scrambling
program (DeCSS) in 1999 and it quickly began circulating on the Internet. As mentioned earlier, the program is considered to be illegal under
the DMCA's "anti-circumvention" clause, and those who are caught using or distributing it may face criminal charges. Industry leaders say this
is necessary to stop the circulation of DeCSS, which they argue is a tool
that facilitates more illegal copying. But supporters of DeCSS claim that
it does not prevent users from copying disks; rather it only limits the
range of machines that DVDs can be played on. One may still make an
illegal copy, they contend, by transferring the contents of one DVD to
another disk without de-scrambling it. Thus they see DeCSS as a tool
that liberates DVDs so Linux users can play them on their own computers, which they consider to be perfectly ethical.

Another technique commonly used by music companies involves the
insertion of an extra security track containing phony data to copyrighted
CDs. When a user tries to play the disc in a computer, it will not work because the computer will keep trying to play the phony track first. Attempting to play the disc might even cause a computer to crash. However, the
disc will still work properly in most standard CD players because they are
not distracted by the phony track; they play only recognizable music codes.

Leaders in the music industry defend this practice by arguing that
computers are the major vehicles for illegal copying. By making millions of discs that can't be played in computers, they hope to significantly curtail the illegal ripping and sharing of music over the Internet.
But the record companies didn't expect that some creative hackers
would figure out a simple way to crack the copy-protection code using,
of all things, a felt-tip marker. In the spring of 2002, news reports indicated that users who blackened the rim of copy-protected discs with
cheap felt-tip pens had broken the code. Others claimed to have made
discs playable by putting tape or a post-it note around the outer edge.[44]

It's not clear if these actions violated the DMCA, but conceivably,
users who cracked the code would have circumvented copy protection
technology, albeit in a very low-tech manner. Which raises some questions: Is cracking anti-copy code ethical? Is telling others how to do it ethical? Is marking up a legally purchased CD with a felt-tip pen the same
as using DeCSS to decrypt a DVD? From a completely different angle, is
it ethical for music companies—or movie studios, for that matter—to
scramble their products so that they can only be played in certain devices?

Looking beyond the actions of the felt-tip hackers and the entertainment industry, there may also be another ethical dilemma in this story. In the process of developing an article about the code breakers, the Reuters news service cracked the security code on a CD itself. After researching Internet reports in which hackers bragged about cracking the code with markers, a Reuters correspondent tested the technique to see if it really worked. The following description of the test was included in the published article:

> Reuters obtained an ordinary copy of Celine Dion's newest release "A New Day Has Come," which comes embedded with Sony's "Key2Audio" technology.
> After an initial attempt to play the disc on a PC resulted in failure, the edge of the shiny side of the disc was blackened out with a felt-tip marker. The second attempt with the marked-up CD played and copied to the hard drive without a hitch.[45]

Was testing the felt-tip method on a protected CD an ethical thing to do? Perhaps the reporter had no other way to determine whether or not the hackers were telling the truth. It appears that he learned that the code had been cracked from messages posted to Internet newsgroups; in such a situation, it would be difficult to authenticate the hackers' claims. Either the reporter would have to physically witness someone marking and playing a disc or he or his co-workers would have to try it themselves. In addition to marking the CD and playing it in a computer, the story indicates that the reporter then copied it "to the hard drive without a hitch." Was it necessary to make what could be construed as an illegal copy of the CD? Does the fact that the reporter described the testing and copying of the disc as part of the story make it ethical?

Many other journalists have committed acts that could be considered illegal in the process of alerting the public to security breaches or exposing a seemingly ridiculous or unenforceable law. Journalistic codes of ethics typically warn against doing anything illegal in the process of news reporting unless all other avenues for gathering important information have been exhausted. Reporters are also advised to explain their actions as part of the story.

Given the previous discussion about digital copyright law and the various criticisms of it, the acts of the Reuters correspondent might seem rather insignificant. But the point in bringing this up is to highlight the fact that most all journalists will find themselves in a position where they are tempted to engage in activities that might seem relatively harmless but, nonetheless, run counter to the law. There is no hard and fast rule for resolving such a dilemma. Ultimately, journalists must carefully consider whether committing the act, or acts, will better serve the interests of the public with minimal harm to other stakeholders.

Copying News Content

In addition to enabling users to efficiently trade music and movie files, the Internet also allows users to circulate news content, to which they may attach their own comments. By simply cutting and pasting an article from an online news service to a bulletin board and inserting his or her own critique of the reporting, a user can attempt to spur an online conversation. Other readers observing the post may then offer their own comments on the article, or on the first poster's critique, and so on and so on—informing, arguing and correcting each other as they go, which is the essence of public debate.

However, it is important to realize that while there is a vast amount of public domain content on the Internet, most news articles are copyrighted just like entertainment products. So the question arises: Is it ethical to make and distribute copies of news content? On the one hand, users could argue that news is a public good and that exchanging it to promote public discourse is, indeed, ethical. After all, circulating copies of news stories offline for noncommercial, educational purposes is considered ethical as well as legal, so why shouldn't doing it online be treated the same?

News organizations, on the other hand, might argue that it costs a lot to produce online news content and that such costs can only be recovered if users read the news on the organizations' Web sites where they will be exposed to advertisements that finance the site. On another level, they may also argue that news content is private property; therefore, circulating free copies of it is the same as pirating movies or music and should be considered both unethical and illegal.

Arguments very much like these were presented in a 1999 court case involving a Web site, called Free Republic, that hosted a forum where users regularly posted, without permission, the full text of news articles from major news Web sites and offered their own comments about the quality of the news reporting, often accusing the news media of having a liberal political bias. The *Los Angeles Times* and the *Washington Post* sued the operators of the site for copyright infringement. The court ruled that even though the site didn't charge user fees, it could be considered a commercial operation because it solicited donations and posted advertisements for other politically oriented sites. As such, the court ruled that posting articles was a commercial use of copyrighted material, which violated the rights of the news organizations.[46]

Presumably, the court's decision leaves open the possibility that noncommercial copying and posting of news articles is a fair use. If so, it would be legal, but would it also be ethical? Professor Litman and others suggest that so long as the reproduction of material doesn't exploit the work commercially, it is ethical. News organizations might counter that all copying without permission is unethical. Resolving this dilemma might turn on the question of whether or not one considers news to be a different kind of product than entertainment or other physical products. C. Edwin Baker, a prominent legal scholar and First Amendment theorist, has argued that news is an important public good, necessary for public discourse and personal enlightenment.[47]

Given Baker's view and considering the fact that news reporting typically involves the gathering of perspectives on events that happen in the public sphere, one could argue that news content should be held in common ownership much like a public park or transportation system. News organizations, however, will likely contend that they specifically, rather than the public at large, have expended time and money to produce the content, thus they've earned the right to control how it is distributed. In a sense, then, cases such as the one involving the Free Republic Web site invite us to consider much broader questions about the ethicality of applying copyright law equally to news and entertainment products. Richard Stallman, as we've noted, has called for differing levels of protection based on the use, or purpose, of a work. Does news have a higher purpose than entertainment? If so, what sort of copyright scheme, if any, would be appropriate?

Trademarks, Domain Names and Other Protected Works

In addition to music, movies, software and news, many other types of intellectual property can be found on the Internet. Some clip art, for example, may be copyrighted, and the unauthorized publishing of it on a Web site might be considered infringement.

Trademarks, which are located somewhere on almost every Web site, are very likely to be protected. And these need not be in the form of bold company logos—any symbol, slogan, sound or name, including domain names, used to distinguish a company and its products from others can be considered a trademark. Even design elements such as colors and shapes that are associated with a business (e.g., the golden arches) can be a type of trademark called "trade dress."

If registered with the U.S. Patent and Trademark Office (PTO), trademarks are protected in a manner similar to copyrights—the owner has an exclusive monopoly on the use of the mark. Web sites with registered trademarks should display the ® or ™ symbols in order to give notice that material is protected. Infringement occurs when a party uses another's registered trademark in a way that causes a "likelihood of confusion" between goods produced by the parties or implies a relationship between the parties. Trademark "dilution" may also occur when the unauthorized use of a "famous" trademark diminishes the identity of the mark or when it is used in a manner that tarnishes the image of the trademark holder.[48]

Web designers and site operators who wish to avoid being pulled into court should be especially careful to determine if design elements they intend to use are similar to, or the same as, protected trademarks.

Aside from the legal issues, there are some ethical concerns too. Building and maintaining a credible Web site, especially for a news organization, requires a great deal of attention to the "look" and navigation of the site as well as proprietary issues. Barb Palser, an online news editor and columnist for the *American Journalism Review*, suggests that ethical, and perhaps legal, conflicts can be more easily resolved if Web developers address the following questions sooner rather than later:

- Do you post copyrighted notices on your own material?
- Do you take logos or images from privately owned sites for use in related stories? If so, do you notify the owner or request permission?
- Do you attribute copyrighted images textually and/or link to the source site?
- Do you use generic images from personal home pages or icon libraries?
- Do you take images from other sites for non-news content?
- Do you modify images taken from other sites?[49]

Palser also recommends that online news editors and Web designers develop a basic understanding of copyright law. As a general rule, it's safe to use images from government sites or corporate logos, so long as they are presented in related news coverage. "The ground gets shaky," writes Palser, "when there could be a perception that the material is being used to earn revenue . . . rather than to enhance a story."[50] Ethically, the best protocol might be to borrow only what's absolutely necessary for clear, factual storytelling.

Linking to Copyrighted Material

As a result of the Free Republic decision just summarized, the operators of numerous Web forums and e-mail lists instituted policies that prohibited the posting of news content in full. If users wanted to comment on a news story, it was suggested that they quote only a few sentences or a paragraph directly and include a hyperlink to the story. To many users and news organizations, this appeared to be a sensible alternative.

However, others questioned the ethics of linking to copyrighted material without permission—could linking be construed as illegal copying? Given that links can connect to either a home page or to internal content, which is where the full news article is likely to reside, should there be one protocol for home page linking and another for "deep linking," which connects users to internal content?

The case study in this chapter will explore this question in more detail; therefore, we'll use this section to consider some broader questions on the subject.

The Purpose of Links

For starters, what is the nature of a link—what does it do, what is its purpose? The most elementary answer is that a link is a connection between two Web pages. Consistent with this definition is the view that a link is something akin to transportation and once clicked, it takes the user to another location, or address, on the Web.

But this isn't the only metaphor used to explain links. Some users and content producers have argued that rather than transporting a user somewhere else, clicking a link actually retrieves or downloads copyrighted material into one's browser. And as such, links facilitate the illegal copying of Web pages. People taking this position often argue that linking without permission is unethical and may even be considered as copyright infringement. Tim Berners-Lee, the computer scientist credited with inventing the Web, rejects this assertion and argues that a link is nothing more than a digital referral or footnote. "The ability to refer to a document," he writes in his book *Weaving the Web*, "is a fundamental right of free speech. Making the reference with a hyperlink is efficient, but changes nothing else."[51]

Linking and the Law

Berners-Lee's position has been accepted by at least one U.S. court, which ruled that linking to both home pages and internal content "does not itself involve a violation of the Copyright Act."[52] Some notable cases in other countries, however, have not come out the same way. In 2002, a Danish court ruled that a commercial news aggregator's practice of deep linking to content on news Web sites violated European copyright law and was a form of unfair competition. Also in 2002, a German court ruled that deep links to database content violated a European Union law that grants copyright protection to database owners.

The courts, it appears, have made the legal issues that much more complicated. On the one hand there is the European trend toward recognizing deep links as a type of infringement while the U.S. courts (at least as of late 2002) have offered little guidance other than to suggest that all forms of linking are legal.

Which brings us back to the ethical questions: Is linking to copyrighted material without permission ethical? How about linking to a home page? Or deep linking? Should there be differing ethical protocols for them?

An Ethical Protocol for Linking?

Hoping to provide some guidance, Richard Spinello, a professor of management and ethics at Boston College, has answered, yes, there is a significant difference between home page links and deep links and that Web publishers should follow different rules for each.[53] Spinello bases his argument on the rationale that while the Web is an open medium, individual Web sites are not common property; they belong to their creators, who, he believes, have the ethical right to exert some control over who links to their sites and how they do so. In this way, he questions the argument that "one's mere presence on the Web is an implied license to link" and suggests that the rights of the user community must be balanced against the rights of Web site creators.

After staking out this position, Spinello points out that Web sites are often built with a specific navigational plan in mind. That is, the designer intends for users to see certain advertisements or artistic features in a particular order, thus they structure pages and links to facilitate a seamless presentation. Deep linking may defeat the creator's right to fair compensation, Spinello argues, by steering users around advertisements, and it may defeat the site's aesthetic presentation—much like a movie theater showing a movie out of order so that the last scene appears first. He also contends that Web site authors "should have the right to restrict links from certain sources when such a connection might be a source of embarrassment." As an example, we might consider a news site that is being linked to by a pornographic or racist Web site. In this situation, Spinello believes the link could give the impression that the targeted site endorses or has some affiliation with the source of the link, which might harm its reputation and credibility.

Spinello suggests that an ethical linking protocol would consist of an open standard for linking to home pages, meaning that one could "assume an implied license to link without permission to any target site's home page." If a Web site subsequently learns that they are the link target of a site the operators find offensive, they should have the right to ask the source site to remove the links. On the question of deep linking, Spinello suggests that in all cases, Web sites wishing to connect to internal content should obtain permission before doing so.

Opposing Views

Spinello's protocol does not sit well with a number of Web users who feel that the medium was deliberately designed to allow users to shift instantly from idea to idea (and page to page) in a nonlinear, nonhierarchical manner. Berners-Lee indicates that it was his intention to develop a medium that made random connections, or *links*, from one subject to another "in an unconstrained, weblike way" that mirrored, as closely as possible, the intuitiveness of the human brain. Hypertext itself, as Berners-Lee notes, was conceived as a language in which the reader "could follow links and delve into the original document from a short quotation."[54]

From this position, attempting to guide users through a series of advertisements or story items they would rather skip defeats the original design of the Web. To use an analogy from another medium, it would be like telling VCR users that they could not fast-forward through commercials or parts of a taped program they find uninteresting. In fact, some leaders in the entertainment industry have argued that the use of so-called "smart" personal video recorders (PVRs), which allow users to record entire programs without commercials, is tantamount to "stealing the program."[55]

Darren Deutschman, an intellectual property consultant who generally supports unrestricted deep linking, suggests that such rhetoric, as well as attempts to reign in linking through the legal system, simply reflect corporate, commercial values, which treat virtual space no differently than physical, or real, space where private lots and businesses are protected by walls, fences and gates. An opposing set of values posits that cyberspace is quite different from physical space in the sense that traditional, commercial borders don't—or shouldn't—exist. The conflict between these two value systems appears to be driving the deep linking debate.

"The deep linking issue," Deutschman told Wired News, "attempts to answer the question that's been asked since the Internet first became part of the general public's consciousness: Is this medium a free source of information for the benefit of the people, or a controlled presentation of branded content that benefits commercial interests?"[56]

Sidebar

Is It Stealing or Just Guessing?

In October 2002, Intentia, a Swedish software company, filed a criminal suit against Reuters PLC, after the wire service linked to the firm's Web site and located and then published the still unreleased third quarter's financial report, apparently by guessing the correct URL where that information was stored. Intentia called the event "an unauthorized entry."

Reuters' spokeswoman replied that "We are rejecting Intentia's allegations completely. Information was accessed from the company's Web site and in the public domain. It wasn't a private site. It wasn't password protected. (The report) was on their public Internet site; it was published, and therefore we reported it."* Reuter's Editor in Chief Geert Linnebank

---

* "Intentia: Reuters Hacked Financial Results," ITWorld.com, October 29, 2002. Retrieved from the World Wide Web: www.itworld.com/Tech/2325/021029reuters/.

weighed in by saying, "Reuters is in the business of informing the market with breaking news stories using all the tools at its disposal, but doing so in a legitimate, ethical manner with journalistic integrity."[†]

This case raises several issues. What constitutes public information and what constitutes private? According to *Forbes* magazine's Web site, Forbes.com, Linnebank said that "This information was in the public domain and available from Intentia's public website where the company itself put it." In this case, it would seem hard to fault Reuters from a legal standpoint, at least.[‡]

But Intentia chose to define "public" from a completely different perspective—that is, it cites Swedish laws regarding "publicly released" information, and according to those laws, claimed Intentia, its new earnings reports were not yet made "publicly available." In this case, Reuters could be accused of accessing private information.

But this case raises larger ethical issues regarding the use of other's information stored on their Web site. Among them:

- Who should determine whether information is considered public or private? Should it be solely up to the creator of the information?
- Is it okay to "guess" URLs to locate information on a public server that the creator did not want to be made public? Which analogy would be more appropriate to describe this activity: (1) trying a bicycle lock combination until it opens and then taking the bicycle, (2) looking through another organization's trash cans for sensitive material that was thrown away and then publishing it, (3) reading and publishing a company's internal memo that you found inadvertently left on a coffee table in an office lobby, or (4) publishing a financial table from a company's financial report that the firm requested you not publish as it was published too early by mistake.
- What party should shoulder the majority of the burden in determining whether information located on a publicly available Web site is considered public or private? Should it matter how the Web page was accessed (i.e., via a search engine, by reference from another site, by guessing the URL)?

All media professionals who use information from the Web, and not just media professionals who work in the online medium, will want to consider

---

[†] Reuters Press Release, October 28, 2002. See http://about.reuters.com/newsreleases/art_28-10-2002_id1093.asp.

[‡] "Reuters Rejects Swedish Company's Allegations," Forbes.com, October 28, 2002, www.forbes.com/markets/newswire/2002/10/28/rtr768765.html.

these matters as part of an ethical standard. A practical consideration to keep in mind in a case like this is what will be the consequences of publishing information that the owner did not mean to make public, and will it be worth it?

Determining Appropriate Levels of Protection

In an act of digital protest against deep-linking restrictions, a man in Chicago created a Weblog that purposely disobeys the policies of dozens of sites that attempt to impose linking rules on other sites that wish to link to them. The blog, called Don't Link to Us!, bears the following notice:

> The Linking Policy for Don't Link to Us! precludes us from requesting permission to link to a site, and compels us to link directly to the targeted page (i.e., a "deep link") rather than to a site's home page.[57]

The site's creator, David Sorkin, is a law professor who hopes that exposing and ridiculing what he calls "stupid linking policies" will convince the Web sites to accept the openness of the medium and rethink their rules.

We might consider the ethics of deliberately resisting the wishes of other Web sites in the manner Sorkin has chosen. But, perhaps, it is just as important to delve into the question of how far Web sites should go in trying to protect their own material? Is trying to impose "no linking" policies a proper way to protect one's intellectual property? How about the restrictions on deep linking?

Many, including Sorkin, have argued that such policies are unenforceable. Widely used sites could not possibly track down every offending linker, and even if they could, it would be virtually impossible to require them all to get either verbal or written consent before linking. Processing all of the requests would take an enormous amount of time and use valuable resources. Rather than attempt to do the impossible, critics suggest that Web operators that really want to protect their property should code their internal pages so that visitors trying to link to them are automatically rerouted to the home page. Others suggest that companies could construct their sites in a way that would require visitors to go through a registration page before accessing internal content. And Web sites that don't want their pages indexed by search engines or news aggregators can use technical means to "hide" their pages.

However, like encryption and other copy-protection techniques, these practices might actually inspire cyberactivists and hackers who reject any restrictions on Web content to devise creative new methods to get around copyright blocks.

Going to Court

Many copyright holders have decided that lawsuits, then, are the best mechanism for protecting their works online. Numerous cases have been brought to the courts by copyright and trademark holders who aim to stamp out infringement. Writing in the magazine *Editor & Publisher* in 2000, journalist Wayne Robins highlighted a few notable cases involving news organizations that took or threatened to take legal action:

- After a Web site called FitnessLink began using the slogan "All the news that's fit," *The New York Times* legal department stepped in and claimed that the slogan ripped off the *Times'* famous "All the news that's fit to print" slogan. FitnessLink decided not to challenge the *Times* and chose a different image line.
- In 1999, Dow Jones & Co., owners of *The Wall Street Journal*, objected when a couple from Maine attempted to register *Small Street Journal*, the name of its 6,000-circulation children's newspaper, as a trademark. Dow Jones' action effectively put Web publishers on notice that it would pursue anyone attempting to register a takeoff on the name of its flagship publication.
- Also in 1999, *The Washington Post* went after the operators of a Web site called *washintonpost.com* (basically the same spelling of the newspaper minus the "g"), which displayed pornographic images. A court agreed that using a nearly identical domain name tarnished the *Post's* highly recognizable trademark.
- In 2000, the *Post*, along with the *New York Times*, the *Los Angeles Times*, and Gannett Co., which publishes *USA Today* and 97 local newspapers, sued a San Diego wireless service provider that had been offering its customers free access to those companies' Web content. A court ruled in favor of the news organizations.[58]

Beyond the news industry, leaders in entertainment and commercial software businesses have made a point of aggressively litigating against those whom they feel are committing copyright infringement. In the same congressional testimony in which he referred to the unauthorized downloading of music and movie files as "digital thievery," Jack Valenti asserted that the courtroom was the "first front" in the battle to protect copyright. "Put simply," he said, "whenever a new site appears whose prime allurement is the illicit availability of movies, illegitimately file-shared or readied for download, it is our intention to move with celerity to bring them to the courtroom."[59] Valenti added that the MPAA was using "a sophisticated search engine" called Ranger "to track down movies illegitimately [placed] on the Web." When Ranger locates an offending site, the MPAA sends "cease and desist" letters to the ISP "whose customer is engaging in the infringing activity or, where possible, to the site itself." Valenti told Congress that in 2001, the MPAA sent ·out 54,000 such letters to over 1,500 ISPs around the world.

Web activists and copyright critics are, no doubt, alarmed by such vigilant efforts to protect proprietary content. They've argued that such behavior doesn't make sense at a time when the movie business, on the whole, is actually generating high profits. Valenti himself acknowledged to Congress that the cultural industries are "responsible for some five percent" of the gross domestic product (GDP) of the United States. In addition, he stated that the "movie industry alone has a surplus balance of trade with every single country in the world."[60]

In such an economic climate, where there are certainly copyright threats but where the entertainment industry also appears to be on strong financial footing, are the protection measures overzealous? Are greed and the desire for control over cultural production driving such policies, as critics have argued? On another level, are company executives and legal departments, rather than artists and authors, pushing for greater control? If one answers yes to these questions, one might also claim that aggressive litigation combined with restrictive linking policies and the use of technological protection devices is unethical.

On this matter, Michael McFarland, a computer scientist, ethics scholar and college president, makes the case that much depends upon the type of intellectual property that is to be protected. In the case of works that have scientific or educational value, such as a new study or medical finding, creators and copyright owners are not being "virtuous" unless they "strive to make the study as available as possible to anyone who would benefit from it."[61] McFarland bases this argument on the view that the purpose of information is communication, and that unless it is communicated and shared, information is "worthless."

As for entertainment works, or products such as commercial software, McFarland admits that determining the appropriate level of protection is "more complex ethically." Nevertheless, he argues, creators and owners must at least attempt to devise distribution schemes "that would provide maximum value to the user community while giving the developers enough compensation to make their labor and investment worthwhile." McFarland suggests that most current laws and protection measures to not achieve this balance. Perhaps, then, as Litman has suggested, commercial exploitation should be vigilantly guarded against and opposed by creators. "Beyond that," McFarland argues, "there should be as free an exchange of information and ideas as possible."

Active Noncompliance with Copyright Laws

No matter what laws Congress passes or how aggressively copyright holders litigate, it appears that vast numbers of people have simply taken the position that they will not comply with intellectual property laws. Professor Litman and other copyright experts believe this is due largely to the overly complex nature of the statutes or the idea that when the laws are deciphered, they just don't make any sense to Internet users. Litman believes that this phenomenon may very likely make the current copyright laws unenforceable and drive the creation of new ones that offer more to the user community, such as explicit fair use exceptions modeled after more traditional copyright regimes. However, given the determination of the cultural industries to push for greater protection and the close relationship between lawmakers and industry, Litman doesn't expect any significant reversal in the trajectory of digital copyright law in the near future.[62]

So where does that leave us in terms of the ethicality of active noncompliance with copyright law? Probably in a short-term phase that will be characterized by widespread "cyberdisobedience" and increased vigilance on the part of industry, though no one can be absolutely sure.

In such an environment, Litman argues that users must determine whether or not they feel as though they are obliged to obey the law because it is the law, or whether they feel it is ethical to obey only those laws that they believe are just. If one decides to pick and choose laws that one will obey, is he or she ethically obliged to accept the consequences if caught breaking the law? Penalties for copyright infringement can range from fines in the thousands of dollars up to awards in the hundreds of thousands of dollars or to imprisonment for up to five years. Litman suggests that there is a strong argument to be made in favor of civil disobedience, though such action will not come without significant costs. For users threatened with lawsuits or criminal penalties, illegally downloading or uploading a few CDs to save a small sum of money may not be worth the trouble.

Moving from the individual scale to the community scale, organized noncompliance efforts are likely to generate greater exposure for the cause of cyberlibertarians and others whose values don't square with efforts to fence in online property. On this issue, John Perry Barlow, a noted cyberspace theorist and former songwriter for the Grateful Dead, suggests that increased vigilance on the part of the cultural industries will continue to have a galvanizing effect on young Internet users who are emerging from a state of political apathy.[63]

This new generation, according to Barlow, values the convenience, interactivity, service and ethics of cyberspace, which, he believes, will promote greater creativity and innovation "following the death of copyright." By ethics, Barlow means the idea that people will be inclined to "reward creative value if it's not too inconvenient to do so." He also rejects the utilitarian philosophy that artists and authors are motivated primarily by monetary reward. Instead, he believes "most genuine artists are motivated primarily by the joys of creation." Not that Barlow thinks they shouldn't be fairly compensated, he simply feels that compensation will be more just once creative works are liberated from antiquated laws and can flow more freely to those who desire them and wish to share them with others. So rather than seeing all forms of file sharing or "piracy" as threats, Barlow sees them as opportunities for artists to get free exposure. As he puts it, "the more a program is pirated, the more likely it is to become a standard."

By linking these strands of thought, Barlow asserts that online services that make the exchange of cultural products more convenient and facilitate peer-to-peer community relationships will quickly be embraced and financially supported by users, while services that erect political and economic barriers around creative works will be swiftly rejected by the masses. Simply put, Barlow doesn't believe that laws tailored to the interests of the physical, industrial economy can prevail in the digital environment. "No law," he writes, "can be successfully imposed on a huge population that does not morally support it and possesses easy means for its invisible evasion."

In Barlow's vision, the trajectory of digital media development will ultimately turn in favor of the rebels, and he suggests that the more the dominant industries push for control, the more the rebels will mobilize against them, speeding up the ethical transformation, or even elimination, of digital copyright law.

Case Study: Protecting Content, Policing Links

In this chapter, our case will once again focus on hyperlinks, but within the context of intellectual property rather than free expression. That is, first, whether sites *may legally prohibit* other sites' linking to their pages and second, whether sites *should* protect against other sites' links and the ethical ramifications of doing so. Because news sites are a sought-after target for others' linking, this issue is particularly germane for the online journalist.

Hypertext is a flexible language that allows Web publishers to connect to Web pages in a number of ways. We can identify three specific types of hyperlinks: (1) links that connect Web users to the home page of a Web site, (2) deep links that bypass the front page and connect users to internal pages or files, and (3) links that insert internal Web pages or files from one site into the frames of another—commonly known as "framing" (Web site operators often use deep links to connect users to content on their own internal pages. As such, links of this variety won't be dealt with in a copyright context).

Linking and Copyright Law

It would also be helpful to know the law surrounding intellectual property and linking. Starting with framing, as of late 2002, the courts have not issued a definitive ruling. However, one notable case, *Washington Post Co. v. Total News, Inc.*, was settled when the defendant agreed to stop framing the plaintiff's content. Attorneys representing *The Washington Post* and five other news organizations that joined the *Post*'s suit, had claimed that the use of hypertext markup language (HTML) frames on the Total News Web site was a violation of copyright law. Rather than pointing visitors directly to the *Post*'s site, the attorneys argued, Total News had used links that, when clicked, appeared to copy the *Post*'s content and insert it into the Total News site, which contained advertisements and the Total News logo. After initially contesting the claim, Total News agreed to drop the frames and link directly to the plaintiffs' sites. The settlement appeared to have put an end to this particular type of framing, and there has also been general agreement among Web publishers that it is unethical.

On the issue of deep linking, the courts appeared to offer some guidance in March 2000 when a U.S. district court judge ruled, in *Ticketmaster Corp., et al. v. Tickets.Com, Inc.*, that "hyperlinking does not itself involve a violation of the Copyright Act." In an analogy to the offline world, the judge argued that using deep links was essentially the same as using "a library's card index to get reference to particular items, albeit faster and more efficient."[64] In other words, copyright law protects the content but not citation pointing to it.

The decision appeared to clear the way for unfettered deep linking; however, it left open the possibility that linking could be construed as unfair competition in situations where the identity of the site being linked to is not clear. Attorney Michael Overing, a specialist in Internet law at the University of Southern California and a columnist for the *Online Journalism Review,* stated that unfair competition claims might also be made when any link, even one to a site's home page, "creates the impression of a business relationship or endorsement between the 'linked-from' site and the 'linked to' site."[65] It's still difficult for a publisher to show proof of damages, making a lawsuit of this type quite risky.

Given the continued growth of the Web and the fundamental role links play in connecting users to content, Overing expects more challenging cases in which the courts are likely to react with "piecemeal" and "inconsistent" rulings rather than definitive judgments. As we await legal guidance, publishers may have even more incentive to work on ethical solutions to the linking dilemma—solutions that are respectful of intellectual property, the design of the medium, and the rights of others in the Web community.

Which brings us to the case, in which a major news organization attempts to protect its content by issuing a strict no linking policy.

The Case: Permission to Link

Since the mid-1990s, NPR has made much of its broadcast programming accessible on the Web. In addition to streaming audio of the latest news programs, visitors to the network's Web site, NPR.org, have free access to a vast archive of program segments dating back to 1996. The sheer volume of material, which also includes articles and transcripts to broadcasts, has helped make NPR.org a popular news and entertainment destination on the Web.

In March of 2002, NPR updated the "Terms of Use" page on its site to highlight its linking policy: "Linking to or framing of any material on this site without the prior written consent of NPR is prohibited . . . "[66] The network made no distinction between links to its home page and deep links to internal content. The policy, in effect, would force every Web site, Weblog and search engine wishing to link to any part of NPR's site to obtain formal, written permission beforehand. The NPR page also included a form that required would-be linkers to list their name, e-mail address, Web site address, street address, phone number, contact information for the person maintaining the link, proposed wording of the link and the accompanying text, and the length of time the link would be active.

Network executives and attorneys argued that the openness of the medium made it easy for others to misuse NPR's content or damage the network's reputation. According to NPR ombudsman Jeffrey Dvorkin, the "proliferation of Web-based activities" made NPR "hesitant about allowing just anybody to put a link to NPR on other Web sites."[67] Specifically, Dvorkin was referring to content aggregators who, NPR claims, were creating commercial Web sites based on links to audio files from NPR.org and other Web sites. Unlike links to text-based Web pages, many deep links to multimedia files leave doubt as to the originator of the content. Typically, clicking a deep link to a specific audio file will open the file in a media player rather than take the Web user to a conventional HTML page with text and graphics.

NPR officials also say the policy was designed to protect their journalistic independence from advocacy groups that place links to NPR stories in such a way that "one cannot tell that NPR is not supporting their cause." "[A]s a journalistic organization," Dvorkin wrote in an online column, "NPR must maintain its status as a disinterested observer and reporter."[68]

Reaction to the linking policy. Few Web users noticed NPR's policy until June 16, 2002, when a Web writer named Cory Doctorow (known by some in the Internet community for his popular Web directory site BoingBoing.com) published a link to the permission form on his Weblog and referred to the policy as "brutally stupid." Almost immediately, the policy was the dominant topic on hundreds of Weblogs, discussion sites and e-mail lists. Many critics argued that NPR had voluntarily published its content on accessible Web pages and had, therefore, implied that others had permission to view or link to it. To require written permission before linking was, they felt, an attempt to circumvent the open architecture of the medium and the libertarian values of the Web community. Moreover, it was also hypocritical, they asserted, because NPR.org regularly deep linked to content on other sites.

Other critics suggested that NPR could easily prevent commercial aggregators from misusing audio files by installing software, or other devices, that would reroute visitors coming from the aggregator's site to the NPR home page. Or they could institute a registration system that would require visitors coming from third-party sites to log in before viewing internal content. If an aggregator broke through those fixes, NPR would have a strong legal case against them. Better to follow that option, the critics contended, rather than trying to limit the rights of honest Web users.

From another angle, Staci Kramer, a columnist for the *Online Journalism Review,* argued that it was "unacceptable" for NPR, a public news outlet that prides itself on being journalistically independent, to solve linking problems "by squashing independence in others."[69] Similarly, Steve Yelvington, an executive with Morris Communications and former newspaper editor, argued that by requiring permission to link, NPR was attempting "to dictate whether and how others may discuss or refer to their Web operations. That's inappropriate for any organization, and reprehensible in a journalism context."[70]

Not all the comments about NPR's policy were critical. University of Illinois journalism professor Eric Meyer indicated that he understood NPR's concern about content misuse and even asked, "What would it hurt to ask someone before you post a link? Not only could you learn whether there might be objections, which you can talk about. You also might learn whether the URL is stable, whether the server could handle the kind of traffic you expect to generate, etc.—things you should be looking out for on behalf of your readers anyway."[71]

NPR changes course. Stung by the sharp criticism, NPR significantly altered its linking policy on June 27, 2002. Instead of requiring that linkers fill out the detailed permission form, network officials posted a message on the site informing visitors that NPR "encourages and permits links to content on NPR Web sites"; however, it clearly states that framing of NPR sites is prohibited. The statement also tells visitors that links to NPR from other sites should not be used in a manner suggesting that NPR endorses their products or causes, and it reminds them that links should not be used for "inappropriate commercial purposes." NPR maintains that it has the right to ban links that violate these two terms. Regarding the use of NPR's logo, the policy warns visitors that it is protected by copyright law and may not be used "without NPR's prior written consent." After the revision was posted, ombudsman Dvorkin wrote that it "more appropriately reflects the libertarian sprit of the World Wide Web."[72]

Many of the critics who had come out strongly against the original policy were quick to applaud NPR for making the revision. Kramer, however, was bothered by the vague words "inappropriate commercial purposes" and by NPR's insistence that it could withdraw linking permission from a site it deemed unacceptable. "I still flinch", she wrote in an *Online Journalism Review* column, "at the notion that NPR might look at a site you or I would consider educational and deem it inappropriate because it doesn't meet NPR's standards."[73] Doctorow also took up the permission issue, suggesting that NPR was actually "failing its commitment to journalistic ethics" by asserting that "permission to link can be extended and revoked" by a Web publisher.[74]

Alternative strategies and suggestions. Clearly, the NPR case highlights a cultural clash within the Web community that won't be easily resolved, especially if there is no constructive dialogue. With this in mind, Meyer suggests that Web publishers use their own interactive medium to explore ethical solutions to the deep-linking dilemma. This could include Web forums to discuss support for, and opposition to, deep linking, as well as an online project to develop voluntary standards for linking. Meyer acknowledges that this would not solve every problem, but "it could start some dialogue that might get people, while not accepting opposing views, to at least appreciate that they exist."

Summary

Copyright is a type of property right that recognizes that the creator of an original published or recorded work has the right to control how that work is produced and distributed. Copyright laws protect this right.

Gutenberg's invention of the printing press in the 1400s made copying documents immensely more efficient. This spurred attempts by the church and European monarchs to censor unauthorized works, which led to a licensing arrangement to permit publication, and this development paved the way to the beginning of copyright laws. Over the next few centuries, the number of books that were published increased enormously. A doctrine eventually evolved that acknowledged that a property right existed between a creator and not just his or her physical properties like land or material goods but intellectual properties as well. And so it became accepted that the producers of books, like owners of physical property, should also have protections.

The English Parliament passed the first modern copyright law in 1710, the Statute of Anne. In the United States, the Constitution recognized intellectual property rights in order to "promote the progress of science and the useful arts," and this led to the first federal copyright law in 1790, which established a copyright term of 14 years. While initially limited to "maps, charts and books," copyright law was eventually extended to other media like music, art and motion pictures, and copyright terms were lengthened. The modern copyright law emerged in 1976. This new version continued the trend of broadening rights for copyright holders and increasing copyright duration, which in 1976 was set at the life of the author plus 50 years. The 1976 act also enshrined the fair use doctrine, which allows users to make and distribute copies of protected works provided they are for "criticism, comment, news reporting, teaching (including multiple copies for classroom use), scholarship or research."

The emergence of the Internet and World Wide Web spurred fears by the major copyright holders (e.g. movie studios, record labels and software manufacturers) that online users would now be able to pirate digital content and easily and instantly distribute it online. In response, Congress passed various laws that expanded copyright protection online, including the DMCA, which prohibits the manufacture, sale, or use of any program designed to circumvent technologies that protect copyrighted works. Some scholars and critics have questioned the expansion of these rights as unfairly and inappropriately shifting the balance between copyright holders and information users to the copyright holders, who in practice are often large media corporations.

Perhaps the most contentious and well-known digital copyright controversy relates to the downloading of copyrighted content without paying for it or obtaining permission. Such an act is seen as an illegal and unethical act of piracy by some and more akin to sharing, similar to the loaning of a book, by others. There are a wide range of views and theories as to what should or should not be permitted to be downloaded and shared on the Internet, and these views often are derived from a particular philosophical point of view about the nature of property, sharing and online. From a practical standpoint, some content firms, particularly music companies, have created technologies to prevent unauthorized downloading by Internet users who either don't know, don't care or purposely want to circumvent the copyright holders' restrictions and actively participate in "cyberdisobedience" as a kind of protest against restrictive copyright protections on the Internet.

Online media professionals face copyright dilemmas in several areas. Among them are the ethics of copying news content, images or other protected information from other sites; whether linking to other sites' news stories (deep linking) could be construed as unauthorized copying; the ethics of accessing information on a Web site that did not intend to make that information publicly available; and prohibiting other sites from linking to one's own.

Critical Thinking Questions

- If you could create a new copyright law, would you give users of intellectual property more or less rights than currently exist, and why?
- If you had a teenage child and found out that he or she was downloading copyrighted music from the Internet, what kind of action would you take, if any? How would you explain the ethics involved to your child?
- Is linking to another news story more like citing it, summarizing it or reprinting it? What are the implications regarding the need to obtain permission to do so?
- If you linked to a story published on another news site and the publisher asked that you remove the link, would you? Why or why not?
- If a Web site with an extremist political view that you found repellent linked to one of your stories, would you request that the site owner remove your link? What argument would you use? What would you do if the site owner refused?
- In reading the NPR case study, did you feel sympathetic to NPR's cause or did you feel that their original regulation was unreasonable? Explain.

- Should the tradition and "spirit" of the Internet of making information freely available impact how you would create a linking policy for your site? Explain.

Key Terms

Berne Convention
Copyright
Copyright Act of 1976
Deep linking
Digital Millennium Copyright Act (DMCA)
Fair use
First sale
Intellectual property
Public domain
Sony Bono Term Extension Act (CTEA)
Statute of Anne
Trademark

Recommended Resources

Books

Bettig, Ronald V. *Copyrighting Culture: The Political Economy of Intellectual Property* (Boulder, CO: Westview Press, 1996).
Lessig, Lawrence. *The Future of Ideas: The Fate of the Commons in a Connected World* (New York: Random House, 2001).
Litman, Jessica. *Digital Copyright* (Amherst, NY: Prometheus, 2001).

Associations and Institutes

Creative Commons. Web site: www.creativecommons.org/.
Stanford Center for Internet and Society. Web site: http://cyberlaw.stanford.edu/.
U.S. Copyright Office: www.loc.gov/copyright/.

PART 3
Journalistic Ethical Dilemmas for Online Media Professionals

In the third part of this text, we examine three specific ethical concerns that are directly related to the practice of journalism over the Internet (online journalism): (1) accurate reporting, (2) quality reporting, and (3) distinguishing advertising from editorial.

We end this section with a separate discussion on the larger role of the Internet and its capacity to further democracy.

To begin this section, it would be useful to recount the emergence and growth of online journalism as a separate medium.

The Growth of Online Journalism

Although online journalism is a relatively new phenomenon, journalists have actually been regular users of computing technology since well before the emergence of the Web. In fact, for many years preceding the Web, a completely separate field of study within journalism existed (and still exists today) called computer assisted reporting or CAR. CAR is a technique that gives journalists tools and skills to utilize computers more effectively in their job.

CAR became popular in the 1980s and early 1990s. During this time, this specialty was devoted primarily to training journalists in how to use PCs for stand-alone uses like searching fee-based databases (e.g., Dialog, LexisNexis, etc.) and using productivity tools like Excel spreadsheets for calculating and displaying data in their published articles.

By the early to mid-1990s, though, CAR training also encompassed using the Internet. CAR instructors taught journalists how to do useful tasks like information gathering and performing research via the Net, how to use the Net to communicate with sources, and how to tap into Internet-based journalism discussion groups.

During this time, though journalists were *utilizing* the Internet as a powerful tool to improve their traditional print or broadcast jobs, they were not practicing what was to eventually be called online journalism. The first instance of online journalism came with the birth of the first Internet-based newspaper—the *Chicago Tribune*'s Chicago Online, which appeared on AOL's service in 1992. At this point, the emergence of a true "fourth kind of journalism, next to print, radio, and television journalism with its own specific journalistic characteristics" could be said to have been born.[1]

The Birth of Online Journalism

Within a very short time, additional news organizations began putting their own Web versions up on the Internet. According to Editor & Publisher's MediaInfo online news tracking site, the number of newspapers appearing on the Web grew dramatically during the early and mid-1990s.

By the mid-1990s, the term *online journalism* was being utilized by some to describe Internet-based news sites, although other names were also being used—"new media" journalism and "digital news," for example. The name "online journalism" eventually stuck and became the accepted terminology.

It is important to note that at this early stage in online journalism's development, there were not yet many software products available that made publishing on the Internet simple. Few journalists were skilled in working in HTML or familiar with the requirements of working on the Internet. Many of these early online news sites, therefore, staffed their operation more with people technically proficient in computers and using the Internet, such as programmers, rather than persons skilled or trained in journalism. New job titles were also created to describe those in charge of putting the news out over the Web, such as "content editor" or "new media producer."

During this early stage of online journalism, there was a great deal of excitement and energy surrounding the capabilities of the Internet as a force for providing news and information. Some observers were enthralled by the idea that the Internet allowed the communication of information directly to the reader, cutting out the media, which for many were viewed as an outdated, unnecessary, biased and elite "middleman." The following is an example of the kinds of comments that reflected this perspective, posted by an amateur but passionate media watcher on a newsgroup in 1993:

> The coming (this is just the warm-up, folks), communications/information revolution will end the concept of an elite minority "collecting" and then "reporting" information to be consumed by the masses. Every individual will be able to post information to the world at large; every other individual will be able to read it, search it, analyze it, discuss it. There's a murder on your block? Tell the world. Digitize the bloody corpse and let a thousand first-graders download it and print it instantly. . . . There will no longer be "opinion makers," those whose job it is to pronounce the "will of the people." There will be no need for people to bind into groups to be heard or be ignored because they are not a "big enough" collective. So

what then becomes of the information professionals, those self-appointed arbiters of the truth, whose job it is to decide for us what news is relevant and what is not; who is a "real" candidate and who is not, which opinions may be expressed in the polls and which may not? What becomes of them? What has become of the village blacksmith?[2]

Others saw this new form of journalism as the wave of the future and that stodgy journalists set in their ways had better adapt—or else. Author Michael Crichton, in a speech to the National Press Club in April 1993, warned journalists that they would need to evolve along with the Internet —or suffer the same fate of extinction as his dinosaurs of Jurassic Park.

Among those who were actually involved in starting up or working at a nascent online news operation, there was a great deal of heady excitement about what it meant to be a journalist in this new medium. An Internet-based "online news" discussion list was formed in 1994. During the mid-1990s, members of this list engaged in lively debate and analysis of their novel challenges, many of which had never before been confronted by traditional journalists.

Some of these early online news pioneers became well-known columnists and experts in the field in their own right. For example, Steve Outing began a column called "Stop the Presses" for the digital version of the trade publication *Editor & Publisher*. This column covered the online news as a separate industry, reporting on what online news sites were doing, how they were evolving, and ways they handled new issues. Other respected voices that emerged from this field included columnist J.D. Lassica and journalism professor Eric Meyer, among others.

Journalists and editors working for online news sites discovered that they were facing issues with no precedents or guidelines. The *San Jose Mercury News* was one of the very first prominent newspapers that began a full-fledged online site called Mercury Center, launched in 1993. One of the authors of this book sat in on an editorial meeting in 1994 and watched as the staff grappled with unprecedented types of decisions and scenarios. One of the editors enthusiastically noted during the meeting that "we're inventing the rules as we go along."

Early Concerns

But while those who were involved in getting this new medium off the ground were energized and excited by these new challenges, there was increasing worry and concern among those who toiled in the traditional media world about the impact of this new medium on their profession.

Traditional journalists, media analysts and others observed that reporting news over the Web could be performed by anyone, without any journalism experience or training, with very little investment, and could potentially reach millions of persons. Furthermore, the Internet itself in the mid-1990s was still very much an unknown quantity, known more for its trivia and lack of organization than as a significant, scholarly or legitimate research or communication tool.

Nobody personified these worries more than Matt Drudge and his Web-based news site the Drudge Report (see www.drudgereport.com). He launched his Drudge Report in April 1995, was profiled in the "Style" section of *The Washington Post* in 1996, and by July 1997 became famous when his site reported that *Newsweek* magazine was "hot on the trail of a woman who claims to have been sexually propositioned by Clinton on Federal property."[3] In January 1998, Drudge's site named Monica Lewinsky and her subpoena, which was then repeated that same day on the Rush Limbaugh radio show.

After becoming famous for breaking this and other stories—some true and others proven to be false—Drudge became a celebrity. He even addressed the National Press Club in 1998, as a representative of his new kind of journalism—Internet reporting that held little respect for the traditional methods of journalistic procedure, particularly as regards accuracy. Drudge has been quoted by New York University's School of Journalism (where he also spoke) as saying, "the reports on my Web gossip sheet are 80 percent accurate" as well as, "I have no editor, I have no fact checker."[4]

It was also around this time that notable errors of fact were being made by companion Web versions of mainstream newspapers. Both the *Dallas Morning News* and the *Wall Street Journal* Online were forced to correct errors in their Web versions (errors that also related to the Clinton-Lewinsky scandal). These cases are discussed in our focus on speed and accuracy online, in chapter 7.

Concern was mounting among traditional journalists about how this new, fast-growing medium was going to impact the profession. By the early and mid-1990s, the journalism profession had already suffered for years from a crisis of credibility. As an institution, surveys showed that journalists were not trusted by the public. In 1997, for instance, Bob Giles, outgoing president of the ASNE, noted the following in a speech to his colleagues about the erosion of trust in the media that was being reported by surveys and polls:

Only 21 percent of respondents in a Wall Street Journal/NBC poll rated the news media very or mostly honest. A Gallup Public Confidence poll in 1996 revealed that only 29 percent of Americans express a great deal or quite a lot of confidence in newspapers. This number is 6 percentage points lower than confidence in television news.[5]

Furthermore, readership of newspapers was dropping. In particular fewer and fewer younger people were getting their news by reading a daily newspaper.

Against this backdrop of declining credibility of the media emerged this new kind of "journalism" and it was causing even more worries. Did the Net's imperative for speed and its sense of immediacy mean the country would see more Matt Drudges—people with no journalism skills, training or inculcation in the profession's ethics? Would individuals simply publish rumors and put out unverified information with no regard for its veracity?

The Profession Takes Action

A frenzy of activity began in the profession to try to figure out what this was going to mean for journalism. Associations and blue ribbon panels were formed to analyze and recommit to the traditional values of the profession, journal articles and books were written that discussed the potential dangers of unchecked online reporting, and experts at professional conferences bemoaned what was happening to the old principles.

Out of this soul searching, several reports and white papers were ultimately published, one notable one being the ASNE 1995 report *Timeless Values: Staying True to Journalistic Principle in the Age of New Media*. Reports like this identified problem areas and then encouraged journalists to affirm and maintain their traditional values during these times of fast and unsettling changes. A 1997 "statement of concern" by prominent journalists led to the creation of the Committee of Concerned Journalists to try to organize their colleagues around the country to discuss their profession's problems and come up with solutions.

In 1999, prominent journalism professors and writers Bill Kovach and Tom Rosenstiel tried to pull all these concerns together and offer solutions in their book *Warp Speed: America in the Age of Mixed Media*.

All the while traditional journalists debated and discussed these concerns, the number of new news sites appearing on the Web continued to explode. The Web site of the media trade publication *Editor & Publisher* tracked the emergence of online media and documented this growth during these early years. According to a U.S. newspaper industry survey conducted for the publication, 62.2 percent of daily newspapers had Web sites in 1997, and by 1999 that percentage had increased to 80.8 percent.

Amid these concerns, in 1999 a new organization was formed, the Online News Association (ONA). The ONA was composed of journalists who work on the Internet or present their information via other digital platforms. According to ONA's Web site, among the association's founding key principles were:

> *Editorial integrity:* "Responsible journalism on the Internet means that the distinction between news and other information must always be clear, so that individuals can readily distinguish independent editorial information from paid promotional information and other non-news";
> *Editorial independence:* "Online journalists should maintain the highest principles of fairness, accuracy, objectivity and responsible independent reporting"; and
> *Journalistic excellence:* "Online journalists should uphold traditional high principles in reporting original news for the Internet and in reviewing and corroborating information from other sources."[6]

A primary reason for the creation of ONA was to enhance the reputation and credibility of the online media. In that vein, in May 2000 the ONA joined up with Columbia University's Graduate School of Journalism and created the annual "Online Journalism Awards" to honor outstanding examples of journalism conducted online. Categories created include General Excellence, Breaking News, Enterprise Journalism (Investigative and Original Reporting), Service Journalism, Feature Journalism, Creative Use, Innovative Use, and Commentary.

A New Era

By the year 2000, the online media industry had evolved quite a bit since its first few years of its existence. By the summer of that year, a survey conducted for *Editor & Publisher* found that all of the 100 largest newspapers were online. And by the fall of 2001, its database listed over 5,000 newspapers on the Web around the globe (3,200 from the United States); over 4,000 magazines worldwide (over 2,600 U.S.);

over 2,200 radio stations (over 1,500 U.S.); and over 1,400 television stations (over 1,200 U.S.).[7]

By the beginning of the new century, there was not just growth in numbers but a greater focus and awareness of the need to ensure quality and recognize traditional journalistic values. And the online media itself was becoming much less a novelty and moving more into the mainstream. In March 2000, according to the Pew Internet & American Life Project, an estimated 52 million Americans, or about 60 percent of Internet users, had gone online for news, and by October 2002, that figure increased to 82 million or about 70 percent of all Internet users (Pew also calculated that by October 2002, 60 percent of all Americans were Internet users).[8]

Another evolution was the increase in the number of familiar media names solidifying their position on the Web. No longer was online news associated with Matt Drudge or an obscure Web-only start-up. Instead, the big names in the online media world quickly began resembling the big names in print and broadcast media: CNN.com, NYTimes.com, USATODAY.com, Fortune.com, *The Wall Street Journal*'s wsjonline.com, and washingtonpost.com, to name a few. These became the well-known brands that Internet users could turn to when they wanted news online.

In fact, they did turn to these big names on the Web. According to a December 2002 Nielson/Net Ratings poll, over half of the top 20 news Web sites in the United States during that month were affiliated with newspapers or well-known broadcast sites. Heading the list was MSNBC.com. Following it were AOL News, CNN General News, Yahoo! News, ABC News, NYTimes.com, Gannett Newspapers and Newspaper Division, Internet Broadcasting Systems Inc., Washingtonpost.com, USATODAY.com, Hearst Newspapers Digital, Time Magazine, McClatchey Newspapers, MSN Slate, World Now, Fox News, The Boston Globe, LA Times, NYP Holdings and Netscape News.[9]

In addition to the nationally known media that launched a site, many hundreds of regional and locally respected names in the news also created their own Web presence. According to journalism professor Eric Meyer's NewsLink, a leading online news site that identifies and summarizes information about online media, as of early 2003, the following major metro newspapers online were the most popular of all U.S. news sites on the Web, as measured by the number of incoming links:

1. Washington Post
2. Los Angeles Times
3. New York Times

4. Miami Herald
5. USA Today
6. New York Post
7. New York News
8. Atlanta Journal-Constitution
9. Dallas Morning News
10. Washington Times
11. Philadelphia Inquirer
12. Boston Globe
13. Chicago Tribune
14. Detroit Free Press
15. Phoenix Arizona Republic
16. San Francisco Chronicle
17. Tampa Tribune
18. Orlando Sentinel
19. Baltimore Sun
20. Charlotte Observer
21. Chicago Sun-Times
22. Cleveland Plain Dealer
23. St. Louis Post-Dispatch
24. Indianapolis Star
25. Fort Lauderdale Sun-Sentinel[10]

Not only were more Americans going to the Web to get their news and more prominent news organizations moving online, but users were finding that their forays onto the Net to get their news were turning out to be quite satisfying. According to the earlier Pew study, most Americans (69 percent) say they expect to find reliable, up-to-date news online (85 percent of Internet users reported this versus 43 percent of non-Internet users). And a full 87 percent of those who reported going to the Net to find news told Pew that they did indeed find what they sought.

So by early in the 21st century, the world of online journalism had grown quite a bit since its emergence in the early and mid-1990s. There were many more sites both from mainstream news organizations and start-ups, and thousands of personal Weblogs as well. And there were many more users—apparently satisfied ones. So what about those early worries that journalism practiced over the Internet was going to retard the profession? Were they proven correct, or did they turn out to be unfounded?

The most dire of the predictions have not come to pass, and as this book will recount, there is a great deal of promise in this new medium. But the jury is still out on this question. And it is a complex question, since the answer also turns on who we choose to call a journalist in the age of the Internet (as discussed in chapter 3).

But much of the answer as to whether the online medium will produce quality journalism will depend on the decisions you make. While some matters in how the medium evolves will be out of your control, you will also have choices to make that will influence its direction. If you are cognizant of the traditional values and ethics of journalism, prepared for the wholly new situations that you'll encounter, and can think through and apply ethical principles to these new challenges, the future of online journalism, and therefore the future of the media in general, will be a more hopeful one.

Recommended Resources

Associations and Institutes

The Poynter Institute has created a timeline to track the growth of the new media from 1969—1998. Web site: www.poynterextra.org/extra/Timeline /index.htm.

Discussion Groups

The "Online-News" discussion group consists of reporters, editors and others who work online or have a close interest in online journalism. The list is hosted by the Poynter Institute. Subscription information is available on its Web site: http://talk.poynter.org/.

CHAPTER 7

Speed and Accuracy

Chapter Goals:

- Examine the role of speed as it may impact the accuracy of online journalism.
- Surface other concerns regarding accuracy of online journalism, including the size and nature of staffing and methods employed for making corrections.

Nearly all U.S. media codes of ethics, as well as international codes, identify *accuracy* as a vital component of ethical journalism. In fact, since misleading and deceiving others, even unintentionally, can cause harm, conveying accurate information is one of the most basic components of all types of ethical communication.

This chapter examines how the practice of online journalism may present challenges to a media professional's ability to comply with the ethical standard of accurate reporting.

Although all journalists need to be vigilant about the accuracy of their work, online journalists may have a special obligation to be particularly careful. The powerful transmission capability of the Internet means that an item posted on the Internet one morning could potentially be read by millions of people around the country or even the world by noon. And since journalists monitor online media for leads, an item posted on a popular new organization's Web site (e.g., USATODAY.com, CNN.com, etc.) will spread the fastest.

From an ethical perspective, what this means is that at least for some online journalists, their power to reach and influence readers can potentially be greater than that of their colleagues in print. If a story is published in print only, and not online, it will be read by the publication's readers and some pass-along readership. But published online, the item's reach may very well extend well beyond the circulation of that individual news site.

Journalism 101

Stepping back a bit first, it's important to note that the importance of verifying information to avoid making errors and inadvertently releasing inaccurate information is a basic tenet of the profession—Journalism 101—and hardly needs justification or explanation.

Unfortunately the entire media seems to fare poorly when it comes to accuracy in its reporting. And it's not a new complaint by any means. In the mid-1970s the prominent pollster George Gallup surveyed people who, at some point in their lives, had been interviewed for a media story. A full one-third of those polled reported that the reporter had gotten the facts wrong.[1] This study would also serve to confirm suspicions of those people who complain that they detect obvious errors when they are knowledgeable about an incident that is then subsequently reported in the press.

While it's hard to pin down precisely just *how* inaccurate the media really is in its handling of facts, there is a variety of potential reasons why a journalist might get a particular story wrong. *Groping for Ethics* author Ron F. Smith outlines several possible reasons why journalists might make mistakes in their reporting:

- Not knowing the community—a lack of understanding of the people being covered
- Carelessness—typos, bad note taking, sloppy research
- Ignorance—not enough preliminary research, difficulty understanding complex matters, especially in technical areas (e.g. math, law, economics, etc.)
- Competitive pressures
- Getting caught up in the story—being overly committed to a flawed report
- The infallibility syndrome—a certainty of being right[2]

So much about the Internet relates to its speed: New information can be put online that's timely up to the minute, and users can access the updated information instantly. In journalism parlance, reporters who work on the Net operate in a "continuous news cycle"—there are no periodic editions that reporters need to wait for in order to enter their latest updated information. That means new information can be disseminated as soon as it is collected.

How does the Internet's capability to be such a fast medium impact the accuracy of online journalism? Speed is not a friend of accuracy. If a news organization's priority is to get information out as fast as possible, critical checks may be bypassed, increasing the chances that bad information will be released.

Indeed, the worry that Internet-based news sites would trade off accuracy for speed has been a concern expressed by the profession since the mid-1990s, when online journalism began gaining momentum and more prominence.

There have been, in fact, certain high-profile cases where an online news site rushed out a story only to later discover that the report was incorrect, resulting in a retraction and correction. One often-cited incident occurred in January 1998 when *The Dallas Morning News'* Web site reported that a federal employee had seen President Clinton and Monica Lewinsky in a "compromising situation" inside the White House, and that that source was reported to have agreed to testify as a government witness. That report turned out to be false, since the source for that information turned out to be unreliable.[3] Interestingly, another case often cited as an example of an online news site pushing out a story too quickly is also related to the Clinton/Lewinsky scandal. An incorrect story on that matter was released by *The Wall Street Journal*'s online site before verification and therefore before the print *Journal* published it in its newspaper. This raised the issue as to whether or not a looser standard for verification was being used online than there was for print.[4]

Are today's online news sites taking less time for verification and fact checking? Are they publishing more inaccuracies than print or broadcast media? We have not located any quantitative studies that have actually tried to tally the number of inaccuracies emanating from online news compared to other media, so this is a difficult question to answer. However there have been studies that examine *users' perceptions and opinions* of the accuracy of reporting in various media.[5] As a result,

most of the analysis of online media accuracy comes from informal observations by those who have observed how the online media is working in practice.

The long-standing concern raised about online news operations is
that there is a greater chance that unverified data will be rushed onto the
site because of several related factors: the Internet's emphasis on speed,
its continuous news cycle, competition from other Internet news sites,
and the pressure to be first with a breaking story.

Upon some reflection, it becomes clear that these concerns are not
unique to the online medium. For example, 24-hour broadcast news
channels also have a continuous news cycle and intense competitive
pressures to break a new story. Another news medium that has a continuous news cycle and is not exactly a "new" media is radio. Wire services
are also competitive with each other, and they too have a continuous
news cycle. (However, wires have a built-in additional accuracy check,
which occurs when a subscribing news source reviews the wire's story
before publishing the story itself.) "The commercial pressure of being
first," says Norbert Specker, CEO of a Zurich-based firm called Interactive Publishing, "will always battle with the editorial pressure of being right."[6]

It does not seem fair, or accurate, to *assume* that online journalism
cannot be as accurate, based solely on the nature of the medium. "I've
seen no evidence of more errors online than in other mediums,"[7] says
Jonathan Dube, a columnist for the Poynter Institute as well as senior
producer for MSNBC.com and editor/publisher of CyberJournalist.net
(www.cyberjournalist.net). In agreement is Eric Meyer, associate professor of journalism at the University of Illinois and managing partner,
NewsLink Associates (www.newslink.org), who says that at least on the
matter of accuracy, while "perhaps not quite up to the standards of print,
with its occasional rather than continuous deadlines, [online journalism]
seems to have come along fine."[8]

In fact, all forms of journalism today function within the same har-
ried, competitive, continually updated media "soup." So while online
news sites should not be singled out as a more likely culprit to provide
bad information, the larger issue is what this hurried, scrolling ticker,
"breaking news alert" type of reporting is doing to journalism as a
whole. As Meyer puts it: "in their haste to breathlessly report news with
a sense of perceived immediacy . . . broadcast, online, wires and even a
lot of print seem to have forgotten that . . . good journalism involves
making judgment, putting developments in perspective and analyzing

potential consequences . . . Ethically, I believe it's not sufficient to fail to make an error. It is incumbent upon [journalists] to tell whole truths . . ."[9]

Online journalism, then, is not *uniquely* susceptible to succumbing to the pressures of competition and being first. But are there any other, special aspects of this medium that might impact its ability to report accurately?

One aspect relates to staffing. The Digital Journalism News Credibility survey, a report released in early 2002 by the Online News Association, reported that of the 72 online news organizations that it surveyed, 75 percent reported having editorial staffs under 10, 56 percent had five employees or fewer, and 25 percent reported having only one online staff member.[10]

Another survey of online news operations, published in the *Newspaper Research Journal,* also found similar small staffs in online news operations. In that study, 46 percent of the daily newspaper online operations polled are staffed by one full-time worker (19 percent) or all part-timers (27 percent). Those surveyed reported an average of six full-time staff members working exclusively for the online operation in one capacity or another.[11]

The problem of small staff size may have gotten worse as a result of the dot com collapse. According to reports cited in the Digital Journalism Credibility survey, many online news sites' staffs were reduced between 10 and 50 percent during the 2000–2001 time period.[12]

A small staff does not, of course, have to mean poor quality control. But if editorial resources are stretched too thin, or there are simply fewer people available to review early drafts of copy, there can be increased chances for inaccuracies. Ron Wolf, chief financial officer of an Internet newswire called Ascribe (www.ascribe.org) notes that stretched editorial staff affects the quality of *all types* of media. "On the newspaper side," he says, "it shows up in more modest ambitions and goals—closing bureaus, shrinking the news hole, undertaking fewer investigations." But he worries that the economics of some online news sites are so severe that it creates a kind of "warped journalism" in that these sites "never have been able to staff their operations in a way that allowed for review of content in the same careful, thorough manner that characterizes a good print newsroom." Says Wolf, "information can't be vetted by the copy desk if there is no copy desk."[13] And, noted the authors of the Digital Journalism Credibility study, "A significantly reduced journalistic staff could have a significant impact on details like spelling, fact checking, grammatical accuracy—and credibility, even if the material is coming from a large, parent organization."[14]

A second aspect of online journalism worth examining is the nature of the people who are creating the online news. What are their skills and training? What is their understanding of core journalistic values? Are the online news sites implementing traditional journalistic ethical codes that focus on ensuring accuracy?

Because it can be simple and inexpensive to disseminate news via the Internet, there are many thousands of online news sites that have been created by single individuals or by shoestring operations with no traditional journalistic background. And, in fact, some of these sites would want to have nothing to do with the established ways of journalism at all. Some of the start-up Internet media sites were created, at least partly, to celebrate a kind of anti-traditional journalism. Creators of these sites view traditional journalism as an exclusionary, arrogant and elite institution that has had too much power in determining what should and should not be designated as "the news."

And it's not just the go-it-alone start-up online sites that wanted to separate themselves from what was seen as the overly stodgy and out-of-touch media. Sometimes even sites established by the larger print and broadcast parents tried to create this same kind of distance. Like other print newspapers that created an online companion site, *The San Francisco Chronicle* made a conscious decision to make its online version, SFGate.com, seem different. Online site editor Vlae Kerschner acknowledged this, explaining that when the site first got started, "the idea then was to hire some people who were not tied to the journalism world, because the Gate's leadership at the time wanted to create a site that was more hip than the newspaper, and stressed unconventional news judgment, counterculture values, and creative headline writing skills."[15]

The proliferation of nontraditional news sites online raises a question. Does the "organization" behind these sites employ anyone who knows or even cares about standard operating procedures for confirming information? Do these sites follow the standard procedures of going back to the original source, using multiple sources, performing internal verification checks, and following other procedures to help ensure accuracy?

According to the Digital Journalism Credibility study, based on its survey of 56 online newspapers and 16 online television news sites, of the total sample, 71.4 percent of the online newspapers surveyed and 83.3 percent of the online TV stations reported that all employees making editorial decisions were trained journalists, either through journalism school, a prior news position or both. On average, 87.3 percent of the editorial staffs in newsrooms that responded to the survey were trained journalists.

According to those interviewed in the study, very early on in the start-up phase of the online news operations, there was a greater need and reliance on technical people such as programmers, and a relatively high percentage of early online news sites were composed of techies, not reporters. But over time, online news sites began realizing that they would be best off if their operations were staffed by journalists, and as a result, the numbers reflect that hiring shift.[16] This study provides some reassurance, then, as to who today staffs the online journalism sites. (Note that the Digital Credibility study did not include Weblogs as part of the online media; we discussed these separately in chapter 3.)

The other reason why there is at least a partial shift to an increase in the number of traditional journalists on online news sites is that more of the news comes from large, traditional news organizations. Wolf says: "the vast amount of news now being displayed on the net is created by AP, New York Times, CNN, Dow Jones, Reuters et al. And it is repurposed for distribution through Internet channels."[17] In fact, most of the popular online news sites today are companion sites that are affiliated with large, well-known major traditional journalistic organizations; for example, *The New York Times* has www.nytimes.com, CNN owns CNN.com, and so on.

Repurposing information by print journalists and broadcasters to their online sites raises concerns over how much new, original reporting is being done by online news sites, a subject we cover in chapter 8. But it also means that those news items on the Internet were likely to have already gone through those organizations' standard journalistic verifications and checks.

Sidebar

Please Sir, May I Have Some More Time?

What do you do when your editor or publisher tells you to hand in an item or story that you feel still needs more checking and verification? The easiest way to resolve this dilemma is to just go along and hand in your story. And sometimes it may seem that you have no alternative.

But often you do have an option, but you need to speak up and assert the importance and reasons why waiting is a better option. Taking this approach means that you are respecting the rights of your readers, the primary parties to whom you have an ethical obligation.

Rather than just agreeing to turn in a piece with data not yet verified, or simply saying you can't make the deadline, you might wish to consider specific reasons to give your editor to give you the extra time. Below are some suggestions to consider. Are there others that you can think of to add to this list?

1. Reputation of your publication
2. A more complete or compelling story
3. Possible dangers of publishing unverified data
4. Your own personal integrity

When you try to negotiate to get the time you need, be sure to suggest alternatives to give the editor some options. Perhaps there is another story that could run in this one's place. Try to be specific about exactly when you feel your additional checks will be completed—provide that time frame and commit to it.

Changing History?

Another issue related to accuracy relates to how online news sites handle story revisions and corrections. In print or broadcast journalism, when there is a need to fix an error or update an earlier story with new information, it is necessary to publish a completely new item (or broadcast a correction). In contrast, an online news site that wants to correct or update an already published item has the option to simply edit that original item and move on.

This certainly raises some ethical concerns. Beau Dure, a Washington, D.C.-based journalist who studied the operations of online news organizations for his M.A. while at Duke University, said that he has become pessimistic about how some online news sites handle inaccuracies. "Little errors," Dure says, "are quickly 'washed away' when the page is revised."[18]

Mark Deuze, an Amsterdam-based journalist, spent several years studying online journalism as practiced in the Netherlands, and authored a scholarly article that reported his findings.[19] Deuze adds another worry about instant online corrections, wondering whether this capability of easily eliminating inaccurate data "can create a shift in responsibilities where the work mentality is 'first we put it online and when it appears to be wrong we take it out' because the users are not aware of it."[20]

What should, in fact, be done when an online news site needs to make a correction? Should it be erased to prevent others from passing it along? One of the hazards of making mistakes on the Web is that an erroneous item can be linked, thereby perpetuating the life of the bad data. In fact, once information is "let loose" on the Net, it can have a life of its own and be circulated for years. But if an item is removed, where is the record that the error ever existed? Should the erroneous report be kept on the site and a correction appended to it? Or perhaps the new correct data should only have a link to the old data, kept off the main site. In any case, it is important that the reader be aware that a change has been made to the story.

Frank Sennett is an online journalist who also runs a site called SlipUp.com, devoted to publicizing errors in online news sites and advocating that these sites do a better job in acknowledging and correcting their errors. Sennett wrote a piece in *Editor & Publisher* on this topic, where he said, "Every news site should provide readers with a prominent link to its corrections page."[21]

Figures 7.1 to 7.3 are examples of an online news site that Sennett feels does an exemplary job in pointing readers to its corrections—the online site of *The Washington Post*, www.washingtonpost.com.

Another option is strikethroughs, as described by Rebecca Blood, referring to the Boing Boing site (Figure 7.4).

Are online news sites simply eliminating the evidence of their mistakes? How are they handling this issue in practice? There is little consistency among online news sites, each treating the matter in their own way.

Note that making digital changes on the fly also has implications for story revisions. Online journalists can write a story, and then if new information comes to light later in the day or several days later, the story can be changed to reflect the new information. A reader coming upon this revised story would not be aware that it reads differently from an earlier iteration.[22]

One of SPJ's codes under the broader principle of "be accountable" states that "journalists should admit mistakes and correct them promptly." How do you think online journalists can make changes to already published stories and comply with the spirit of this guideline? What would be considered full disclosure to the readers? Would it include identifying changes made for fixing minor errors like typos?

Figure 7.1 The corrections link, while small in this image, is clearly viewable on the left bar of the front page of the site (halfway down left column left of the hand-drawn arrow). Reprinted with permission.

Some Advantages Online

The online medium may actually offer certain advantages when it comes to accuracy. Because the Internet allows for such easy and instant interactivity, online news sites typically set up forums where readers can discuss, debate and hash out the day's stories, either amongst themselves or with one of the online editors.

Some readers take advantage of this feature to point out errors they've noticed in a recently published article. This can result in a quick correction by the online news site, if so required. This self-correcting mechanism of the Internet is familiar to anyone who has participated in an online discussion group. If someone posts something that is incorrect, typically another participant or several participants will quickly— often within minutes—post a rebuttal or correction.

Sometimes the attentive reader who points out an error will even offer a supporting citation to correct or challenge the bad data. This al-

Figure 7.2 Clicking on the link brings up a page of the most recent corrections. Reprinted with permission.

lows editors to immediately check on that reader's claim of error and make any necessary corrections—right away if need be. This "reader check" helps ensure that bad data does not remain on the site for any length of time.

Perhaps it is worth considering that there may be nothing *inherent* to journalism conducted online that would make it more difficult for its reporters to produce accurate work. Rather, the critical factors for any news operation, in any medium, would be to have implemented appropriate fact checking and verification procedures and for its parent organization to have made the budgetary commitments to provide sufficient staffing to carry out these functions.

Summary

A fundamental journalism ethic is the need to verify information and to be accurate. The media's reputation in fulfilling this obligation has

Figure 7.3 Clicking on the top correction brings up the story along with the correction notice posted next to it. Reprinted with permission.

not, unfortunately, been very good over time. And the emergence of online journalism has spurred new concerns that online reporters will do their work hastily and bypass traditional accuracy checks before publishing information as a result of working in a medium that has a continuous news cycle and emphasizes speed.

However, all forms of media today face much of the same pressures as online journalists—a shorter news cycle, intense competition and pressure to move quickly. Reporting online, then, should not and has not yet proven to be inherently more subject to publishing bad information than other mediums.

At the same time, there are certain aspects of online journalism that could negatively impact its ability to check facts and verify information. One factor is staffing, as many online news operations are very small and may not have enough people to do the necessary checking. Another is the nature and values of the people who staff online news operations —the question has been raised as to whether they have the training and the inclination to do the kind of traditional fact checking and confirmation work that is considered de rigueur for traditional journalists. Recent research on this question is reassuring, though, as it indicates that al-

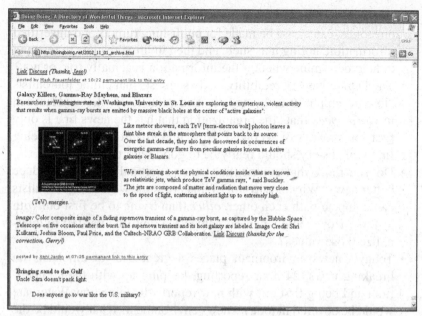

Figure 7.4 An example of a strikethrough form of correcting material, on the BoingBoing.net site. Reprinted with permission.

though the early staffing of nascent online news operations consisted largely of technical persons not trained in the craft, today's online news sites are largely staffed by trained journalists.

A related issue is how an online news organization handles revisions and corrections. The online medium provides the opportunity to digitally erase and alter existing copy. This raises ethical concerns, as such changes would not be apparent to readers. While online news sites vary greatly in their corrections policies, some specific practices and policies have been recommended. The key recommendation is that any changes or corrections be clear to the reader. Specific approaches include clearly posting a link to a corrections page or, where appropriate, putting strikethrough marks through deleted copy.

Critical Thinking Questions

- Do you think that there are factors that would make online journalists more prone to disseminating inaccurate data? If so, what would be some ways around those problem areas?

- Because readers of online news will often respond and comment on stories, can a case be made for publishing rumors and unverified information, noting it as such, and then inviting readers to respond to help determine whether the information is actually true or not?
- The Digital News Credibility study suggests that online journalists "lose the emphasis on being first" and quote Rosenstiel and Kovach in *Warp Speed* that "the organization that has the news first is only first for a matter of seconds." How important do you think being first with a story should really be to good reporters?
- Do you think that competition among journalists helps produce better news? Why or why not? What are some ways that journalists can compete with each other, *other* than trying to be first out with a story? How could a news organization motivate its staff to internalize those values?
- Today's news environment places a great deal of emphasis on breaking news, 24-hour reporting, keeping up with the competition, and being first out with new reports. Imagine how Watergate would be covered in today's news environment. What would be the main differences in how the news about Watergate would be reported? Would this style of reporting better serve the public in understanding the event or not, and why?

Key Terms

Interactive
News cycle
Repurpose
Verification

Recommended Resources

Books

Downie, Leonard Jr., and Robert Kaiser. *The News about the News: American Journalism in Peril* (New York: Knopf, 2002).

Kovach, Bill, and Tom Rosenstiel. *The Elements of Journalism: What Newspeople Should Know and the Public Should Expect* (New York: Crown, 2001).
Smith, Ron F. *Groping for Ethics in Journalism*, 4th ed. (Ames: Iowa State Press, 1999).

Reports

Online News Association, Digital Journalism Credibility Study, January 2002. Web site: www.onlinenewsassociation.org/Programs/credibility_study .pdf.

CHAPTER 8

Sources and Searches: Does the Internet Make Journalists Lazy?

Chapter Goals:

- Identify how the Internet could impact the quality of reporting.
- Discuss the reliability and trustworthiness of information sources on the Web.
- Provide strategies for evaluating information obtained from the Web.
- Examine whether online information discourages original reporting.

The ethical journalist performs quality work. The Society of Professional Journalists places its own efforts toward advancing quality journalism directly at the heart of its mission. It states: "The Society of Professional Journalists is dedicated to quality, responsible journalism as the foundation of a free and informed society."[1]

What is quality reporting? It's not something that can be strictly pinned down by enumerating a list of dos and don'ts. Ultimately quality work becomes clear in its final form: You know it when you see it.

But while it is a fruitless task to try to enumerate all the ingredients that produce quality work, it's not so hard to identify one necessary factor. It takes *effort* to produce an excellent report. Lazy reporting will not produce high-quality work, and doing poor-quality work does not advance the higher mission of the profession, therefore it is unethical. Leonard Downie Jr. and Robert Kaiser, whose book *The News About the News: American Journalism in Peril* laments the decline in serious journalism in the United States, succinctly put the matter this way: "Quality is hard."[2]

Two Temptations

In this chapter we explore whether the Internet may pose certain temptations for journalists to put out less effort when producing their work. There are two major factors that we examine here:

Dilemma A: The impact of the ease of locating information by searching the Internet.

Dilemma B: The impact of the ease of locating already written news accounts by searching the Internet.

In both cases, there is the potential for the journalist to not go the extra mile to ensure quality in his or her work. In the first case, there is the concern over whether the journalist who uses information found on the Internet will confirm it for reliability and accuracy. In the second case, there is the concern over whether, in researching a story, the journalist will go beyond what was already written on the topic and do his or her own original reporting. This chapter looks at both of these matters.

An important point that needs stating up front is that this chapter examines certain hazards related to using the Internet to perform research. But it's not just online journalists who are using the Internet to do their information gathering. Most reporters today, whether they work in print, broadcast or online, regularly turn to the Internet to perform research. (See, for example, *Survey of Media in the Wired World*, by Middleberg & Associates. This survey of 4,000 journalists conducted in the year 2000 found that most reporters used the Internet to find new sources, experts, press releases, article research and more.[3]) So although this topic goes beyond the practice of online journalism, we are giving it special attention because this text focuses on those hazards that emerge when the Internet and journalism intersect. But keep in mind that what we discuss in this chapter can be applied today to all journalists.

What the Middleberg survey reinforces is that the Internet has become a mainstream, if not the preferred, source to use to perform research. And there is no turning back from this—there is simply too much useful, valuable information that can be found on the Web—and it can be done in a matter of seconds.

Dilemma A: The Ease of Locating Information on the Internet

The Web has become the place to turn to for almost any kind of research. Part of the reason is that so much good and useful information has migrated to the Web. Another reason is that it is fast, and many sources are free. Another reason is that it's so easy—having access at one's desktop means just toggling to your browser and entering a few words into Google or another search engine of your choice.

Of course, just because searching the Net is the easiest way to do research doesn't mean it's the best. And of particular concern to the journalist is whether the information that is located on the Web is going to be accurate and trustworthy.

Concerns regarding the reliability of Internet information have been a topic of popular discussion for some time. There have been many anecdotal reports of incorrect and deceptive data found on the Web, through Internet discussion groups, and particularly circulating via e-mail. There has been a tendency among some people to look down upon information found on the Internet as less worthy or credible.

Yet savvy Internet researchers, such as librarians, market researchers, investigative reporters, research scientists and others, know that condemning the Internet as an information source is an oversimplification. While inaccurate and misleading information can certainly be found on the Internet (as it can be of course found elsewhere), skilled researchers have learned how to pull together their own collection of trusted sources on the Web and have developed techniques and strategies to evaluate unknown sites.

Even the general public now sees the Internet as a mainstream source of news and information, and, apparently, does not have much problem with what they find. According to a December 2002 study released by the Pew Center's studies from its ongoing series called the Internet & American Life Project,

> the Internet has become a mainstream information tool. Its popularity and dependability have raised all Americans' expectations about the information and services available online. When they are thinking about

health care information, services from government agencies, news, and commerce, about two-thirds of all Americans say that they expect to be able to find such information on the Web. Internet users are more likely than non-users to have high expectations of what will be available online, and yet even 40% of people who are not Internet users say they expect the Web to have information and services in these essential online arenas.

Internet users are very likely to say that they expect the Web to be a source of information on health care, government agencies, news, and shopping. About 80% of Internet users say they expect the Web to have information in these topic areas. These high expectations are driven by experience. *Of Internet users who have sought information from the Web on these topics, about three-fourths have had positive experiences in finding what they need.*[4] (Emphasis added)

What are we to make of this? One could certainly view these findings not as any kind of testament to the quality of information on the Internet but to the naïveté of the public, and some of that view would certainly be accurate. At the same time, we have all by now experienced countless anecdotal reports from friends and colleagues who have found very important and useful information on the Net, whether it be health related or for one's work, school or other day-to-day research needs.

It's not accurate to use a broad brush and dismiss information on the Net as uncredible. Journalists know this, and the Middleberg study referenced previously is illustrative of the integral way the Internet has become part of the practice of the craft. Furthermore, there has been no evidence that journalists (print, broadcast or online) are having an ongoing, systematic problem with bad data gleaned from the Internet.

Still, there have been a few high-profile cases where journalists did indeed use inaccurate data from the Internet, and there are important lessons to be learned from these cases.

For example, there was the case that occurred in August 2000 where a fraudulent press release was circulated through the Internet on a news wire service that reported that a computer company called Emulex was being investigated by the Securities and Exchange Commission (SEC) and that its CEO had resigned. This press release was picked up by several of the major news sites, including CNBC, CBS Market Watch and Bloomberg.

Reporting this inaccurate information did real damage to certain parties. The company itself suffered, and some individual investors lost money when Emulex's stock dropped sharply. Furthermore, the credibility of the news organizations that reported this false story was damaged. (Many news organizations were fooled because the report came through on a recognized news wire, called InternetWire; however, an ex-employee of that service was able to get access to the wire to send the false report.)

There have also been instances where hoaxsters have created Web pages to appear like a well-known legitimate news or company Web site, and then via e-mails or other techniques have directed unsuspected Internet users to that site. One site, Spoofed.com, even had on its site something called the "CNN Fake News Generator" to help people create a CNN-like page before it was forced to take it down by the media firm.

But while there's nothing to support a worry that cases like this have become commonplace, they do occur from time to time, and the Middleberg study did raise a couple of cautionary flags regarding how reporters are checking what they find on the parts of the Internet that are most notorious for rumors and unchecked data, the chats and discussion area. For example, only 44 percent of respondents stated that they would *not* consider using postings from Web chats or Usenet newsgroups as primary or secondary sources for an article. About one-third said they would use such postings if confirmed by another source. And 47 percent said they had reported or would consider reporting information that "started on the Internet," though it would need to be confirmed by an independent source (22 percent would do so if the information was from reliable professional news organizations' Web sites and 9 percent would but would report it as rumor. The study provided a ranking of how much trust reporters put into various types of sites on the Web).

This raises the question as to whether journalists should be receiving training in searching and evaluating Web sites, so that they can do a better job in evaluating sources they turn up on the Internet. According to the Middleberg study, just 27 percent of respondents said that they received training in computer-assisted reporting (which includes all uses of computers in reporting, not just research.) Of those who reported that they did receive training, almost half received it at their own initiative and not directly from their employers.[5]

Perhaps you already feel you are skilled in Internet searching and don't need additional help. Many reporters apparently do feel that they are proficient in this area. A very interesting research project called "News Libraries in Crisis" was conducted by the Institute for New Media Studies, which is part of the University of Minnesota's School of Journalism and Mass Communication. The project convened a panel in the fall of 2001 to examine the role of news librarians and how their roles were shifting within their organizations.[6]

As part of this project, a survey was conducted of both reporters and news librarians (as well as other parties) and portions of the results were quite revealing. One question asked of reporters was to judge their own competence in Web searching; a similar question was asked of news librarians, who were asked to evaluate the competence of reporters in Web searching. As it turned out, 64 percent of reporters said they were either competent or very competent in searching, while only 14 percent of librarians said that they thought that reporters were competent or very competent. Figure 8.1 shows the actual results.

While it's hard to say whether reporters are as good on the Web as they say they are, librarians are in a good position to make a judgment. While reporters, particularly the younger ones, are likely to be very comfortable and savvy in searching the Web, librarians are specifically trained to understand the entire spectrum of online research. This includes not just using a search engine but when and how to use more sophisticated fee-based online services like Dialog or LexisNexis, how to construct advanced Boolean and field searches to make them as precise as possible, and understanding what it is and is not possible to find on the Internet. Librarians understand the complete panorama of research and information sources, including print and other sources not online. They have been trained in how to judge the quality of a particular information source.

Of course good reporters have developed a lot of these skills along the way as well. And so both parties are in a position to learn from each other. For this reason, journalists who want to ensure that their information search and evaluation skills are the best they can be may wish to consider making the effort to obtain instruction from their in-house librarians or from other outside information experts.

There are several options available for journalists who wish to enhance their Web search and evaluation skills. For example, some media firms make in-house training available by the librarians. Some news librarians do even more than train. For example, the news librarians at the Florida *Sun-Sentinel* have pulled together recommended, trustworthy Web sites in a variety of topical areas and put them on the organization's Intranet. When a reporter needs a good Web source for the topic, he or she just opens up that category and can obtain the listing. For example, under "Aviation Disasters," the reporter views sites of the National Air Traffic Controllers Association, the International Airline Pilots Association, the FAA Aviation Safety Data, the National Transportation Safety Board (NTSB) and others.

News Librarians:

On a scale of 1 to 5, 1 being "not competent" and 5 being "very

competent", how competent do you think journalists in general are in

performing information searches without news library assistance?

		Number of Responses	Response Ratio
not competent	1.	1	1%
	2.	27	18%
	3.	98	67%
	4.	16	11%
very competent	5.	4	3%
	Total	146	100%

Reporters:

On a scale of 1 to 5, 1 being "not competent" and 5 being "very

competent", how competent are you in searching for information without

assistance?

		Number of Responses	Response Ratio
not competent	1.	0	0%
	2.	3	2%
	3.	42	34%
	4.	49	40%
very competent	5.	30	24%
	Total	124	100%

(You can view the entire survey on the Web at:

www.inms.umn.edu/research/newslib/Survlinks.htm)

Figure 8.1 A survey created by the Institute of New Media Studies (Minneapolis, MN) found that when asked to rate their online search competency, journalists give themselves higher rankings than their news librarians. Reprinted with permission.

There are also several books and publications, as well as outside conferences and training sessions where you can go and learn valuable Web research and evaluation skills (see the listing at the end of this chapter).

Because this is an ethics text and not a how-to book on searching and evaluating information on the Internet, we will not go into any detail on how to assess the reliability of sites on the Internet. We have, though, taken this opportunity to excerpt some material from one of the authors' recent books, which is about how to evaluate information on the Web. We also list several recommended sources on this topic at the end of the chapter.

Tips on Evaluating Information on the Web[7]

Experienced searchers say that there are clearly some red flags that serve as immediate indicators of whether a site is more or less likely to be trustworthy. These red flags include:

Is it a "personal" page?

Sites that do not have their own Web address are subject to the highest scrutiny. Often these are personal home pages, identified with a tilde ($\sim$) in the address, or by a GeoCities, AOL, or other consumer ISP URL. These pages may represent a single individual that has not made a serious commitment to establishing credibility on the Web, and as such are not given much either.

A lack of contact information on the site.

This problem was mentioned most often. If the site does not say how the individual or organization that created the site can be contacted, this is a very bad sign indeed. While it may not necessarily indicate that there is anything devious or suspicious about the site, at a minimum, it shows a lack of any kind of attempt towards establishing credibility.

Email contact only.

Just a notch up from no contact information are sites that show no physical address or telephone, and only an email contact. This is not reassuring at all, and again shows a disregard towards assuring the user that there is a legitimate, real person or organization behind this site that has an existence beyond the Internet. The worst kind of email only contact is one that doesn't even have the person's name or title, but just a generic one, like Webmaster@xyz.com.

Spelling and grammatical errors.

If a site displays poor writing and spelling errors, it is a clear sign that the creator is either poorly educated, or at least is quite sloppy and careless —neither will inspire any kind of confidence in the quality of the information provided on the site.

No evidence of recent activity.

A site that has not been updated recently is not going to be of use for any time sensitive research; furthermore, it is another example of a lack of attention to a site's maintenance. Sometimes it is possible to get some clues on when the site was most recently updated. For example, does the page list older area codes that have since been updated; or names of company executives no longer in their position?

Also, if you are on a company's home page, you can often view a listing of the most recent press releases (usually found under the "about us," "recent news" or "investor relations" link). If you see a long string of press releases that begin in 1997, but ended 18 months ago, you are right to wonder why nothing more frequent has been posted. Does it mean that there hasn't been any more company news? Does it mean that the firm has opened a newer site and you are looking at an older one left up on the Web? Does it mean that the firm has gone out of business? Whatever the answer, the lack of recent news is unsettling.

Outdated look and feel.

A site that displays no graphics and mid-1990s style HTML heads and body text is indicative of a site that has not kept up with changes, and, therefore, the data on the site may be stale as well. This is not a hard and fast rule, though, as there are some sites on the Web that do contain good information, but have a poor or old-fashioned design. But when you encounter a site that *looks* old, just be sure to do the checks to make sure the *data* isn't old.

Outlandish or peculiar claims.

If a Web site contains information that is so strange that it just doesn't sound believable, it is going to raise the suspicions of any good researcher. You are in the best position to detect this if you are already somewhat versed in the subject you are researching. But common sense plays a role too. Michael Bass, the Director of Research for the Associated Press, recounts such an example as when a Washington State University newspaper published an article in commemoration of Filipino American history month that casually noted that Filipino immigrants arrived in 1587 on the California shores on a Spanish galleon called *Nuestra Senora de Buena Esperanza*, which loosely translates to "The Big Ass

Spanish Boat." The writer lifted that translation from a source on the Web that was a satirical site. A few moments of reflection would result in the awareness that this would be a ludicrous, and literally unbelievable, name for a Spanish boat of the 1500s.

Reassuring Signs

On the other side, seeing the following on a Web site was reassuring and inspired some level of confidence to consider the source worth considering:

Complete contact information.

The best sites provide extensive contact information: not just an email, but a physical address, phone and fax. It is easy to find (either on the home page or immediately visible under a "contact us" link) and lists actual people's names along with their phone numbers and/or emails.

An "About Us" link.

The sites that inspire the most confidence allow the visitor to immediately learn something about who is behind that site, typically via an "about us" link. Here the site's visitors can find out about the origin of the entity behind the site, its mission and agenda, other people involved and their backgrounds, any publications or reports issued, where it's been in the news, its affiliations and associations, and more.

Elegant, intuitive design.

While a pretty site certainly does not mean an accurate site, attention to detail in graphic design and navigation can be an indicator of careful attention to what makes a site "work," a sign of quality and savvyness, and a good indicator that this is a sharp and quality entity behind the scenes, and deserving of some attention. Chris Barton, the Senior Editor of Hoovers Online, says that "if the site is well designed, I'm much more comfortable with getting information and trusting this site; this is based on my experience—it takes a lot of organization to collect and update information and you won't have that sense of organization and presentation for poor sites. So if it is amateurish, sloppy or rife with copy editing errors or difficult to use—it makes me skeptical."

Indication of timeliness.

One of the biggest frustrations about information found on the Internet is the common lack of any kind of date or indicator of the freshness of the page. The best sites will indicate when the site was last updated, and be clear as to what was updated. Because dates can be hard to decipher on the Web (e.g. does "Updated Jan 1 2003" mean that a new piece of information was added? That some designs were changed? That the whole thing was redone, or what?). For this reason, a site that contains references or links to items recently in the news can be reassuring as these are clear indications of recent activity.

If you know something about the topic that the site is covering, you should also be able to scan it to see if recent important developments or news are covered or mentioned in some way. If they are not, it may raise your suspicions. For instance, when I examine pages that discuss effective use of search engines, sometimes I will find some that do not even mention Google at all—in that case it's a pretty good sign that the page was written a few years ago, before Google came upon the scene, and has not been updated; and I will move on to another site on the Web.

Backs up claims.

Patrick Ross, an award-winning investigative journalist who lives in Northern Virginia, says that the sites he likes best are those that not only present their own information and views, but also include links and backup to primary source documents, so he can then go and look at these documents himself and do any additional verification.

Philosophies, approaches, methodologies outlined.

The best sites don't just give you information, but explain why it was collected, methodologies or approaches used in collecting information, and identify assumptions and any limitations to the research. The form that this takes will of course depend on the nature of the site, but the idea is that these sites help reassure you and provide you with an understanding about what went on behind the scenes a bit, and before the data was published.

Good journalists know that when they come across a source of uncertain veracity—no matter where that source was found—that the prudent thing to do is to get at least one, if not two, confirming sources. Another check is to run that suspect information by a colleague or someone knowledgeable in a position to comment on its likely credibility.

If these options are not available, and the credibility of the information remains uncertain, the best strategy is to sit on it and wait to see if a more reliable source or confirmation emerges.

In summary, there's no turning back from doing research on the Web. But it takes some time and practice to learn how to do it effectively, particularly in vetting Internet sources for their accuracy and credibility. It's incumbent upon today's journalists to ensure they do whatever they can to enhance their abilities to evaluate the sources that they use from the Internet and know which types to skip, which can be trusted, and which require additional confirmation.

Dilemma B: The Ease of Locating Already Written News Accounts on the Internet

Reporting Not Repurposing

Quality reporting emerges from several factors, but as all good journalists know, to produce a top-quality, compelling and important story requires sleuthing, digging and hard work. In practice, this means following paper trails to locate and examine original source documents and tracking down knowledgeable trustworthy human sources and asking them probing questions. All of these activities take time, hard work, and, to use the craft's vernacular, lots of "shoe leather."

When online journalism began growing in the early and mid-1990s, some worried about whether high-quality original reporting would be part of this new medium. Some of the online news sites that were initially launched by existing print and broadcast media served primarily as simply another means of distributing the same content, but through the Web. The news was simply, to use the rather unappealing industry term for this activity, "repurposed" (another unattractive term to describe news obtained online in the same manner is *shovelware*).

There were other early worries too. Steve Weinberg, author of *The Reporter's Handbook* and a past executive director of the Investigative Reporters and Editors, Inc. (IRE) association, voiced some of these concerns in an article he wrote for the *Columbia Journalism Review* in 1996 titled: "Can 'Content-Providers' Be Investigative Journalists? The Worries of a Veteran Reporter." In that piece Weinberg wrote that he was "skeptical, about cyberspace journalists being given the time to do original investigative reporting." Weinberg tempered his worries somewhat, though, quoting another reporter who was more optimistic about the future of online journalism. Weinberg said that he hoped that this other, more optimistic viewpoint would win out.[8]

Several years have passed since Weinberg expressed those concerns, so we might now ask how online journalists are doing in their ability to create new, original, quality—and even investigative—reports.

On the one hand, some online news sites still primarily repurpose from their parent organization, and time and budget are still major constraints on the ability of online news operations to engage in much original and investigative reporting (time and budget are factors that also constrain the capabilities of print and broadcast journalism as well). But as of 2003, online journalism appears to be showing promise and potential in expanding into the more inspiring areas of reporting.

There have, in fact, been several examples of high-quality original and investigative reports that have emerged from a variety of independent online journalism sites. Some independent online news sites that have garnered a particularly strong reputation for their online investigations include Salon, Slate, APBNews.com, TheStreet.com, TomPaine.com, CNET and the Center for Public Integrity.

Bill Allison is the managing editor of the Center for Public Integrity. The organization's mission "is to provide the American people with the findings of our investigations and analyses of public service, government accountability and ethics related issues." And it uses its Web site to publish the results of its research projects. Among the investigative series that the center has put on its site: a yearlong study of party contributions and state expenditures; a government program that helps schools and libraries connect to the Internet falling short on oversight and "riddled with fraud"; and "The Buying of the President 2000." Ironically, Weinberg, the author of the *Columbia Journalism Review* article cited previously, is himself deeply engaged in working on a three-year investigative report for the center (called "The Business of War").

Allison says that the staff at the center follows all the normal guidelines of strict investigative reporting, such as "meticulous fact checking." But what Allison says has really made the Web such a useful medium to report its findings is the opportunity to make so many original source documents—many of these obtained by the center via Freedom of Information Act (FOIA) requests from Washington—available directly to its readers.

Indeed, although online journalism may still not yet be taking advantage of the possibilities that are open to it to pursue original and investigative reporting, there are reasons to believe why there may in fact be *greater* opportunities for online journalists to do more of this kind of quality, rewarding work. In fact, participants at a three-day event held by the Poynter Institute with the IRE to discuss investigative reporting and online journalism noted the following reasons why this kind of reporting may be ideally suited to Web-based news operations:

- Traditional print and broadcast journalism, for a variety of reasons, such as the cost to mount investigative work, have cut back on these endeavors. This opens up an opportunity for online journalism to fill the gap.
- Getting a reputation for investigative work helps build a "brand name" for a new online news site.
- Online news sites can provide continuous, ongoing coverage of a topic—what has been termed *issues-based coverage* as opposed to *"events-based coverage."*
- Multimedia capabilities: online news sites can use audio, video, mapping and links to greatly enhance the power and effectiveness of their presentation.
- There are no space constraints—whatever original documents and sources that are located can be posted, no matter the length.[9]

Whether original and investigative reporting comes to pass on a larger scale online is still to be determined. But there are encouraging signs. One of these has been the creation of an online journalism awards event, created by the Online News Association and the Columbia University Graduate School of Journalism, to recognize outstanding reporting by online news sites (see Figure 8.2). The annual awards were initiated in 2000, and the judges award honors in a variety of categories, which include original and investigative reports. Interestingly, the entries are split into two major categories: "affiliated" sites, which are those online news sites that are part of a larger media operation; and "independent" sites, which originated on the Web and have no counterpart in another medium.

An interesting question to consider when discussing how online journalism reporting may evolve is to consider how online journalists view their role. Do they see themselves as having the same or a different mission than traditional journalists? A study by Ann M. Brill uncovered some interesting findings.[10]

Brill's study first cited previous research that indicated that there were indeed correlations between the particular medium that journalists work in and how they viewed their role. Specifically, Brill cited a study by David Weaver and G. Cleveland Wilhoit (that also drew on previous studies) that outlined three different ways journalists saw their primary function: as serving an *interpretive* function, a *dissemination* function and an *adversarial* function. Those authors eventually added a fourth function called *"populist mobilizer"* when they revisited their work in 1996.[11]

Figure 8.2 Daniel Forbes, a freelance reporter, won the 2002 Online News Association award for investigative reporting for this story he wrote for the online journal Salon, on how the White House was quietly working with networks to integrate its anti-drug messages into prime-time programming. Reprinted with permission.

In Brill's study, fewer online journalists than print saw themselves as serving an interpretive or adversarial function. A higher percentage of online journalists saw themselves as carrying out certain specific tasks under the populist mobilizer function category—letting the public express views, entertaining and setting the political agenda.

If the online journalist sees himself or herself more as a facilitator for mobilizing the public, and less as an interpreter of events or in an adversarial position, how might he or she approach investigative reporting? Would there be different goals and priorities?

What about Weblogs, which we discussed in chapter 3? Although the quality and substance of these vary wildly, there's no question that the bloggers do original reporting. Many even do a short form of investigative-type work: identifying intriguing events that may have been overlooked by the mainstream media, making connections of related stories in the news, and analyzing larger social issues. However, a problem in pursuing traditional investigative reporting on the Web is that these kinds of projects normally results in lengthy pieces, which does not neatly fit into the short-chunk-style writing format of the Weblog.

Can Research Be *Too* Easy?

The Internet has made it tantalizingly easy to instantly uncover previously published news stories on virtually any topic. Typing a few words into a general search engine, or a specialized news aggregator, can, depending on the topic, return dozens, hundreds or even thousands of previously published newspaper and newswire stories that populate the Web. (News aggregators are a subset of general Internet search engines and do not search the entire Web, but instead search only a closed set of news sources. Examples of these aggregators include Moreover.com, DayPop, Yahoo! News and the separate news search functions of standard search engines, such as Google News and AlltheWeb's News Search.)

Of course it's true that journalists and other information professionals have been using online sources for many years, primarily via fee-based proprietary services like Dialog or LexisNexis. However, there were certain limitations that held back their use. For one, the databases themselves only provided access to a limited set of newspapers—some of those indexed only a couple dozen major dailies, while the biggest files might include up to a hundred or so. (Fee-based *journal* article databases—like ABI/Inform, Magazine Index and ProQuest, are much larger, often indexing items published in hundreds or even thousands of specialized magazines and journals.)

Furthermore, many of these databases only provided abstracts, as opposed to the full text of the published articles. They were also generally not available from the reporter's desktop but at centralized locations, such as the news library. And the fact that these databases were generally expensive to search was a disincentive for reporters to use them in an unlimited casual fashion. (Some news organizations' libraries purchased these databases on CD-ROM, which were not billed by the hour like online databases were. However, the problem with searching news on CD-ROMs was that these were not current, updated no more than on a monthly basis).

Today on the Internet, in contrast, reporters can instantly browse and search the full text of articles published in several thousand newspapers, newswires, journals and Weblogs from around the globe—and at no cost.[12] So it has become incredibly easy for journalists to sit down in front of their PC, and in a matter of literally seconds, turn up dozens or more published articles and essays that cover whatever topic is being researched.

What is wrong with that? Make no mistake about it—this instant information retrieval capability of the Internet should and does fall under the category of a "good thing." There's no doubt that the ability to quickly turn up name brand, trusted published news (along with, of course, all the other kinds of information on the Web) has been a boon to researchers of all kinds, and journalists in particular. The Internet has provided instant access to information that would once have been considered too arcane, obscure, narrow or published too far from home to be able to track down, or at least do so easily. Few media professionals would want to give up this treasure trove, and as the Middleberg study shows, reporters have continued to flock to the Internet to perform their research.

Information on the Internet has, in fact, been an aid to journalists in several ways. Mark Ingebretsen, a free-lancer who writes "The Daily Scan" column for *The Wall Street Journal* Online, uses the Web to locate links to primary sources that are referenced in the print news articles he's come across in his readings. Gary Price, an author and librarian by training who has his own Weblog for Internet researchers (www.resourceshelf.com), also says that he uses the Web to track down and read original documents, such as a government document or company report, after reading about its existence from a press release or other brief mention in a media source. And although data on the Internet is often maligned as being untrustworthy, savvy researchers who know how to evaluate information online even use their preferred sites on the Web to authenticate and confirm *other* data.

Yet, although the Internet offers much, it also may be taking back something in return. Neil Postman, a New York University professor and noted author, is a technology skeptic who believes that each new technology deals us a Faustian bargain.

What might journalists be giving up as a result of using the Internet for research? Could it be that when it's so easy to surf and scrape together other journalists' work that it will become less likely for a journalist to do the hard work of creating a fresh story, finding his or her own sources and doing new interviews?

Another concern is whether access to all these articles causes journalists to do less of their own questioning and probing. Eric Meyer, an associate professor of journalism in the College of Communications at the University of Illinois at Urbana-Champaign, worries that "special interests are now using the Web to 'spin' issues their own way—to create their own arrays of fact-stacked information—and what journalists are doing, rather than evaluating information, is simply pointing readers to the various arrays in the worst of 'he said, she said' journalistic approaches. This is at once lazy and arrogant. And it does little either to further the societal obligations of the press or to further its long-term potential as a necessary part of information exploration in the world."[13]

There is even the increased temptation that a reporter could simply lift portions of a previously published article from the Web and do no original work whatsoever. The authors of the Middleberg survey noted the "questionable ethical practice" of "the reluctance of many journalists to give credit to other publications' work when using their material from the Net."[14]

The Googlization of Research?

Everyone seems to have taken to Google. But approaching research with a Google mindset can be hazardous. On the one hand, it is widely accepted that Google does, indeed, do an uncannily good job in retrieving relevant results as a result of its algorithm and link analysis methods. And getting back relevant results is the most important feature for an Internet search engine.

And Google has introduced other features that have proven to be extremely useful to researchers, such as indexing PDF, Word and other non-HTML files; including a link to a cached page; providing results in context with keywords bolded; various advanced and Boolean search capabilities; and a clean, uncluttered interface.

But as useful as this incredible search engine is for research, like any tool, Google still has its limitations. For one, although its unique method of ranking pages is acknowledged to work extremely well, in practice it ends up penalizing newer Web pages since these are less likely than older ones to have many other pages linking to them. (This is a simplification of Google's PageRank mechanism, which examines many factors in making a ranking: Pages that have many other pages linking to them are given a boost in Google's rankings, and it is more likely that older pages will get this boost than newer ones.)

Secondly, like all search engines, Google is only capable of indexing a portion of what is on the Web. Much of what cannot be retrieved is part of what is called "The Invisible Web." These are pages that are not retrievable by search engines, either because they are part of password-protected sites, are generated by forms, or simply have not been found by a search engine.[15]

But the biggest problem in relying on Google is the same as the problem of relying on any other Internet search engine: Although you may think you are doing comprehensive research, you are not. The Internet does not contain anywhere near the extent of the world's information. Most knowledge still resides primarily in places that are outside of cyberspace: in books, older print sources and, of course, in people's minds.

Google, other search engines and even the Internet itself are best viewed as an incredibly useful resource for journalists but still just a single resource and not the beginning and end of all research. Many savvy reporters often use the Internet not as an end unto itself but as a means to an end; for example, for locating leads and finding experts and knowledgeable people worth interviewing.

The E-Mail Interview: It's Fast, but Is It Best?

Good journalists know that the kinds of communication and experience one gets from an interview depends on whether the discussion was done in person, by telephone or via e-mail. Each mode has its own pros and cons.

E-mail interviews have become extremely popular. Indeed, even back in 2000, the Middleberg study reported that "E-mail has now matched the telephone as the preferred method for interviewing new sources; e-mail was slightly behind the telephone in the 1999 survey."[16]

E-mail interviews offer certain advantages. One is the ability to contact people that may be hard to track down on the phone or in person. Another is that this format allows both the journalist and the interviewee to compose their thoughts carefully and precisely ahead of time. And there is usually a very fast turnaround time.

But there are some potential downsides to the e-mail interview. The useful and sometimes telling voice cues that occur during an in-person or phone interview are missing. Also, e-mail conversations are static. One person writes something or asks a question; some time later the other responds. Because there is no continuous unbroken back-and-forth discussion flow, a reporter's ability to challenge, question and probe what the subject is saying is more stilted, less frequent and therefore less effective.

Furthermore, it may be misleading to readers to describe an e-mail exchange as an "interview," as that term may still conjure up a dialogue of two persons speaking, either in person or by the telephone. Some news organizations explicitly identify e-mail interviews as such, which prevents any misunderstanding by their readers.

Sidebar

Critical Thinking Sidebar: Google News

Google made quite a splash in the Internet world when it introduced "Google News", an Internet news service laid out like a traditional print newspaper, with images, headlines and blurbs of what Google's algorithm "decided" should be the top news stories for any particular moment (see Figure 8.3).

Google creates this page automatically by using its existing page-ranking algorithm and applying it to over 4,000 news sites on the Web. (Google defines a news source site as one that is "periodically publishing on timely events.") The site itself is refreshed every few minutes. The result is an image like the one reprinted here from January 24, 2003.

When Google first introduced this feature, it added a tagline, somewhat in jest, that Google's news was created "untouched by human editors." But editors and reporters noted that, of course, human editing was performed —it was simply done by the editors of those news sites before Google indexed them. Perhaps as a result of these journalists' observations, Google eventually replaced that disclaimer with one that read, "The selection and placement of stories on this page were determined automatically by a computer program."

Google News is not only useful to get a quick snapshot of the news but also a nice way journalists can see what stories other news organizations, worldwide, are covering, and *how* they are covering them. Google News makes it very easy to compare how dozens or sometimes hundreds of online media sites are covering the same story. Journalists can even trace the history of a developing issue by clicking the "sort by date" function, which arranges the stories in chronological order with the most recent item placed first.

Figure 8.3 Google News selects news stories to highlight and display based on an algorithm and not human editors. Reprinted with permission.

Does this site create any ethical dilemmas? What would it mean for reporters to begin relying on Google News to determine the "top news" and worthy of coverage? Are there assumptions about what Google does or does not include as a news source that need to be considered? What about the implications of the use of a mathematical algorithm to determine what should be the lead story?

Summary

The ethical journalist produces high-quality work, and creating quality takes a great deal of effort. Because it is so easy to do research online, the Internet may tempt journalists to neglect the hard work of digging and producing original reports.

One temptation relates to the use of online source material. Although an enormous amount of valuable, important information resides on the Web, there is also fraudulent, unreliable and bad data as well. It behooves journalists to not just be skeptical searchers but to have specific strategies for evaluating and assessing what they locate online. This knowledge can be obtained in several ways—by taking in-house training classes, by attending professional training workshops, or by reading books and instructional Web sites that provide tips on evaluating sites and data gleaned online.

Another danger of relying on the Web for research is the ease in accessing already-published reports and neglecting the critical though tedious hard work of following paper trails, inspecting original documents, and tracking down and interviewing human sources. There were early worries, in fact, that online journalism sites would merely "repurpose" their own existing news stories and only use the Web as another dissemination medium. Fortunately, although some online news sites do indeed repurpose, other online news sites are breaking new ground in doing original and even investigative reporting. Many believe that online news sites may actually be well positioned to be leaders in quality, original reporting since online news sites can provide readers with access to original documents; provide ongoing, continuous coverage of an issue; and utilize multimedia. Furthermore online news sites can fill the hole left by print and broadcasting media that have turned away from hard original reporting.

There is no discounting, however, that the Internet—particularly the much-admired and popular Google search engine—makes it tantalizingly easy to locate and piece together previously published articles, making some reporters less inclined to perform the hard work of tracking down their own sources. It's important for journalists to realize, though, that as effective as Google is in retrieving useful and relevant Web pages, like all search engines and Internet finder tools, it has its limitations and certainly cannot retrieve all of the various types and formats of knowledge that exist.

Finally, although reporters have found e-mail interviews to be a fast and efficient way to conduct interviews, the value and advantages of the in-person and telephone interview should not be forgotten.

Critical Thinking Questions

- When you perform research on the Internet, how do you determine which sites to trust? What steps do you take when you come across information on the Web that you could use for your research but aren't sure if you can believe that source?

- If you have a librarian in your organization or at your school, in what ways do you think he or she could help you do your Internet research better?
- Think about some feature story or investigative report you recently read or viewed that you felt was outstanding. Describe exactly what made the story so compelling and of high quality.
- How would you approach writing a long, investigative report for publication on an online news site? How would you format or break up the report to meet the needs of the medium and its readers?
- Do you prefer to do interviews in person, on the phone or by e-mail? Why? What are the benefits and downsides of each?

Key Terms

FOIA
Investigative reporting
Invisible Web
News aggregator
Original reporting
Repurposing

Recommended Resources

Books

Berkman, Robert I. *The Skeptical Business Searcher; The Information Advisor's Guide to Evaluating Web Data, Sites, Sources* (Medford, NJ: Information Today, 2003).

Bruzzese, Lee, Brant Houston, and Steve Weinberg. *The Reporters Handbook,* 4th ed. (Columbia: Investigative Reporters and Editors, Missouri School of Journalism).

Mintz, Anne. *Web of Deception* (Medford, NJ: Information Today, 2002).

Sherman, Chris, and Gary Price. *The Invisible Web* (Medford, NJ: Information Today 2001).

Reports

Middleberg & Associates, *Survey of Media in the Wired World*. Web site: www
.middleberg.com/toolsforsuccess/fulloverview.cfm.
University of Minnesota, School of Journalism and Mass Communication, Institute for New Media Studies. *News Libraries in Crisis*. Web site: www
.inms.umn.edu/convenings/newslibraryincrisis/main.htm.

Internet Research Workshops and Training Institutes

National Institute for Computer Assisted Reporting (NICAR). Investigative Re-
porters and Editors. Web site: www.nicar.org/.
The Poynter Institute Seminars. Web site: www.poynter.org/seminar/.
Web Search University. Web site: www.websearchu.com.

Internet Web Site Evaluation Checklists:

Evaluating Internet Research Sources, created by Robert Harris, retired professor of English at Southern California College, Costa Mesa, Calif. Web
site: www.virtualsalt.com.
Evaluating Internet Resources, Elizabeth E. Kirk, Electronic and Distance Ed-
ucation Librarian, Milton S. Eisenhower Library, Johns Hopkins Univer-
sity, Baltimore, Md. Web site: www.library.jhu.edu/elp/useit/evaluate/.
Evaluating Quality on the Net, Hope N. Tillman, Director of Libraries, Babson
College, Babson Park, Mass. Web site: www.hopetillman.com/findqual
.html.
Evaluating Web Pages: Techniques to Apply and Questions to Ask, University of
California, Berkeley, Library. Web site: www.lib.berkeley.edu/TeachingLib/
Guides/Internet/Evaluate.html.
Evaluating Web Resources, Jan Alexander and Marsha Ann Tate, Wolfgram
Memorial Library, Widener University, Chester, Pa. Web site: www2
.widener.edu/Wolfgram-Memorial-Library/ webevaluation/webeval.htm.
*The Virtual Chase: Evaluating the Quality of Information Found on the Inter-
net*, by Genie Tyburski of the law firm Ballard Spahr Andrews & Inger-
soll LLP. Web site: www.virtualchase.com/quality/.

CHAPTER 9

When Business and Ethics Collide: Advertising, the Internet and Editorial Independence

Chapter Goals:

- Review the origins and reasons for the "wall" between advertising and editorial.
- Examine whether the wall has, in practice, been more followed or breached.
- Explore specific ethical dilemmas confronting online journalists and the separation between editorial and advertising: advertising formats that may blur distinctions, such as sponsored content; and the appropriateness of linking to advertisers.
- Discuss actual cases of how online news organizations handled these kinds of ethical dilemmas.
- Identify recommended guidelines created for online media for distinguishing advertising from editorial.

Among the many ethical issues discussed in this book, perhaps none causes more agony and stirs more debate among journalists than the relationship between editorial and advertising departments. On this subject, numerous questions arise, such as: What, if any, contact should reporters and editors have with the sales and marketing staffs of their company? Should reporters and editors be involved in the planning and implementation of advertising campaigns? Conversely, should sales and marketing personnel be involved in the planning and implementation of news coverage? What sort of guidelines should govern the placement and "look" of online advertisements? Who should establish and enforce any such guidelines—news or advertising departments? And how should journalists handle stories that deal with products and services offered by their parent company or stories that deal with the business interests of their bosses and sponsors?

Some journalists adhere to the long-standing belief that the business and editorial sides should be separated by an impenetrable "wall" or strict division much like that which separates church and state in many constitutional democracies. These journalists worry that unless the ethical wall is upheld advertising departments will gradually assume control over areas that were once the domain of editors and reporters. Under pressure to generate large profits, it is feared that news organizations will use the Web to seamlessly integrate interactive advertising into news content, further blurring the line between the two and damaging journalistic credibility.

Others argue that as technology and media economics have changed, so too must newsroom practices. From this perspective, it is unrealistic to think that financial concerns can, or should, be completely separated from the editorial process. Better to have an open exchange between business people and news people so that new revenue models can be developed while, at the same time, the worst abuses are guarded against. Proponents of this approach argue that news editors should be allowed to review, and possibly veto, online advertising packages that might blur the lines between commercial messages and news. Likewise, sales personnel should be able to participate in strategic planning sessions with top editors so that they can organize advertising and marketing campaigns around upcoming news coverage. This model treats both journalists and sales personnel as allies in the common project to make their news organizations highly profitable.

Still others argue that the traditional wall between news and commerce was always more myth than reality, a rhetorical device developed by media owners in the early 1900s in an attempt to silence critics and media reformers who complained about the media's increasing dependence on advertising for revenue and the growth of consolidated newspaper and magazine chains. Many in this camp suggest that, aside from some exceptions here and there, advertisers and media owners had cozy relationships long before the Web arrived. In addition, they argue that as the primary financers of journalism, advertisers have long had the power to influence the content of the news—power that critics contend has increased in recent decades with a new wave of media consolidation and the convergence of media formats (text, video and audio) on the Web.

In this chapter, we'll discuss these arguments in a bit more detail by looking back at significant changes in the structure of the media, both technological and economical, which first brought concerns about advertising and editorial conflicts to the surface. This historical perspective should enable the reader to better grasp the difficulties that arise in the management of advertising-supported news operations in the digital age.

Origins of "The Wall" between News and Advertising

As mentioned in chapter 2, newspaper publishing was not considered to be a profitable business until about the middle of the 19th century. Until that time, most prominent papers were subsidized by wealthy publishers or contributions from major political parties or smaller politically oriented groups (feminists, laborers and abolitionists, for example) who wanted a public platform for their views. The penny presses were an exception as they began to derive more of their subsidy (about 40 to 50 percent) from advertisers, which enabled them to sell their papers cheaply. This began the trend of greater reliance on advertising for revenue—the advertiser replaced the political boss as "patron" of the press.

By the end of the 1800s, the partisan and penny presses had been displaced by a more centralized, highly profitable corporate newspaper system that followed the logic of the commercial marketplace. Newspapers became less overtly political, filling their pages instead with more news about crime, the courts, accidents, high society and leisure activities. This style of journalism was less likely to upset readers with strong political affiliations, thus it helped attract wider, often upscale, audiences, which could be used to lure advertisers eager to promote new mass-produced products. In 1920, advertising accounted for two-thirds of all newspaper and magazine revenue in the United States, and by the 1990s advertising made up roughly 80 percent of print media revenues and nearly 100 percent of broadcast revenues.[1] Realizing the enormous profit potential of the advertising-based revenue model, publishers changed the editorial/advertising content ratio of their papers. In 1940 the average daily newspaper contained about 60 percent editorial content and 40 percent ads. By 1980, the average newspaper carried 65 percent ads and 35 percent editorial content.[2]

At the same time advertising was becoming the primary source of revenue, press ownership was becoming more consolidated. Leading publishers such as Joseph Pulitzer and William Randolph Hearst established national newspaper chains and purchased national magazines, through which they amassed great wealth and political influence. Ownership consolidation led some to question the direction in which the press was moving, namely whether or not the powerful new media moguls could possibly offer reliable reporting given their increased dependence on corporate advertisers and their affinity for pushing news coverage in directions that favored their own economic and political interests.

The solution adopted by many Progressive Era reformers and publishers was not to break up the emerging press monopolies, support unions for journalists, or finance the development of noncommercial papers that would be operated as public trusts; rather they decided to launch professional training programs in colleges and universities, which would emphasize journalists' ethical responsibility to serve the community and teach them to report the news in an unbiased and "objective" manner. Publishers also committed themselves to separating the business and editorial departments of their papers and magazines in order to prevent advertisers or company business interests from influencing news coverage. It has often been reported that Robert McCormick, publisher of the *Chicago Tribune*, built separate elevators for his editorial and advertising staffs.[3]

Though the words "wall" and "church-and-state" have repeatedly been used to denote the ethical separation between editorial and advertising, it is not clear who was the first to put those words in a journalistic context. According to Thomas Leonard, a journalism professor at the University of California at Berkeley, there is no record of McCormick ever referring to a "wall" or sacred barrier between news and advertising, nor is there any evidence that Henry R. Luce, founder of *Time* magazine, ever used the terms. In fact, Luce, who is often credited with being an ethical pioneer in journalism, admitted that he would sell "a small fraction" of editorial independence "for a price." Both he and McCormick—separate elevators be damned—also used their publications to promote their conservative, anti-communist political views. Perhaps, then, the "church-state wall" is, in the words of Leonard, "an advertising slogan for the generation of McCormick and Luce more than it is cold reporting of their words."[4] Whatever the case, the phrase caught on and journalists continue, rightly or wrongly, to use it today.

Advertising and Codes of Ethics

Reinforcing the notion that journalists should be independent from advertisers and other outside influences, the ASNE adopted the "Canons of Journalism" in 1923. Often referred to as the first ethics code for journalists, the canons stated, "The right of a newspaper to attract and hold readers is restricted by nothing but considerations of public welfare." Reporters and editors were also told that "independence from all obligations except that of fidelity to the public interest is vital to maintaining credibility." Again, there is no specific mention of a "wall" or distinct separation between editorial and advertising departments, but the point was clearly made that the content of news should be guided by the public interest and that journalists should steer clear of entanglements with advertisers and other powerful interests. Much of the language from the original canons forms the backbone of the ASNE's current "Statement of Principles," which was adopted in 1975. In addition to calling for editorial independence, the updated code states that journalists "must be vigilant against all who would exploit the press for selfish purposes" and that they "must avoid impropriety and the appearance of impropriety as well as any conflict of interest or the appearance of conflict. They should neither accept anything nor pursue any activity that might compromise or seem to compromise their integrity."[5]

Other contemporary ethics codes have similarly implored journalists to avoid conflicts of interest with sponsors. The SPJ code of ethics states, "Journalists should . . . avoid conflicts of interest, real or perceived" and "deny favored treatment to advertisers and special interests and resist their pressure to influence news coverage."[6] The RTNDA also highlights editorial independence as a major virtue and instructs electronic journalists to

Gather and report news without fear or favor, and vigorously resist undue influence from any outside forces, including advertisers, sources, story subjects, powerful individuals, and special interest groups . . . Determine news content solely through editorial judgment and not as the result of outside influence . . . Recognize that sponsorship of the news will not be used in any way to determine, restrict, or manipulate content . . . Refuse to allow the interests of ownership or management to influence news judgment and content inappropriately.[7]

While these ethics codes deal with conflicts of interest between journalists and advertisers, they don't deal specifically with the design, placement and labeling of advertisements. In the print world, individual newspapers or chains tended to draw up guidelines for ad placement. The standard practice was to insert ads only on the internal pages of a paper or magazine and make them visually distinguishable from news articles.

The Communications Act of 1934, which established the FCC, set the framework for regulating the public airwaves and required that paid commercials be easily identifiable. "All matter broadcast by any station," the act stated, for which money is exchanged shall "be announced as paid for or furnished by" the sponsor.[8] This was necessary during the 1930s through the 1950s when commercial messages were often broadcast live by announcers, including newscasters, during programs. Sponsors would also routinely buy an entire hour or half-hour of programming and advertise their products exclusively throughout the show. Later broadcast stations would insert "breaks" into programs during which prerecorded commercials, usually 30 seconds in length, would be aired. This made it easier to distinguish ads from news and other programming. In cases where broadcasters allowed outside interests, such as political parties, businesses, and public relations firms, to furnish talent, scripts, or other materials for use in public affairs programs, the 1934 act required broadcasters to make an announcement "at the beginning and conclusion" of the programs, which identified the materials and services provided as well as their source.[9]

Through the use of both voluntary ethics codes aimed at professional behavior and commercial broadcasting guidelines, several generations of journalists were instructed to put the interests of the public first, which meant keeping their focus on gathering the news and reporting it accurately, while sales and marketing personnel dealt with advertisers and other business considerations. But how well did these codes and guidelines work through the middle and latter parts of the 20th century? Were publishers and broadcasters able to successfully resist pressure from advertisers? Were they successful at blocking the efforts of their powerful owners to guide news coverage? Did they take special care to keep ads from looking or sounding like news stories and vice versa? It would be impossible to surface conclusive answers—at least quantita- tively—however, we can look at some historical evidence, which sug- gests that while professionalism increased and the look and sound of advertisements became more distinguished from the news, business in- terests continued to play a significant part in shaping editorial content.

Journalism and Business Pressure: Separating Myth from Reality

On one level, the ethical separation of editorial and advertising de- partments may have given journalists some protection and autonomy from owners and sponsors, who may have simultaneously become less heavy-handed in their dealings with journalists. In other words, many of the most obvious abuses that angered progressive era reformers were curbed, but it does not appear that they ever vanished completely.

Direct Censorship

In the 1930s, for example, maverick journalist and media critic George Seldes highlighted numerous examples where major sponsors were able to get editors to kill stories that did not advance their interests or that portrayed them in a bad light.[10] Typically they did this by mak- ing personal contact with owners or high-ranking editors and urging them to back off the story—orders usually proceeded down the corpo- rate chain of command to editors who risked losing their jobs if they re- sisted. In other cases, sponsors punished newspapers after the fact by pulling their advertising. Many publishers learned the lesson quickly: Don't tangle with the people who provide your profits.

Some advertisers have gone so far as to draw up content guidelines for any show in which their ads appear. In the 1960s, an executive with Procter & Gamble (P&G) testified before the FCC that his company es- tablished the following guidelines for television programs:

> Where it seems fitting, the characters in Procter & Gamble dramas should reflect recognition and acceptance of the world situation in their thoughts and actions, although in dealing with war, our writers should minimize the 'horror' aspects. The writers should be guided by the fact

that any scene that contributes negatively to public morale is not acceptable. Men in uniform shall not be cast as heavy villains or portrayed as
engaging in any criminal activity.[11]

P&G had similar rules for the portrayal of businesspeople: "If a businessman is cast in the role of villain, it must be made clear that he is not
typical but is much despised by his fellow businessmen." The company
also mandated that programs should go out of their way to mention industries that partnered with or carried P&G products, such as "the grocery and drug business."[12] The executive testifying before the FCC said
the company's guidelines applied to news and public affairs programs
and documentaries as well as entertainment programs.

P&G was clearly trying to set the context in which its advertisements
appeared. TV commercials are surrounded by programming, thus an advertiser spending millions of dollars on a campaign to promote a product is wary of the information, ideas and personalities that frame the ads.
Naturally, advertisers will attempt to control the context as much as possible. A gourmet restaurant, for example, may not want its ads to appear
alongside stories about world hunger and starvation; therefore, restaurant executives might suggest that a media outlet reduce its coverage of
issues related to hunger and poverty. This is fairly understandable, and
many might even say it is acceptable given the pressure on businesses to
sell products. But what is not so understandable or acceptable is the cooperation of media companies in helping sponsors get control over programming content, especially when it comes to news. While there may
have been many TV stations that resisted P&G's demands, it is worth
noting that the company has for decades consistently been the largest advertiser on television. It is doubtful that they would have kept spending
so much money on commercials without getting something in return.

Self-Censorship

Direct censorship of the sort practiced by P&G and other major advertisers can lead to what Seldes and many others call "self-censorship,"
which occurs when a news outlet decides, without being asked, not to
pursue a particular story or subject. It may also significantly edit or bury
a report for fear that a sponsor, or even an entire industry, might be offended. In what is perhaps one of the most egregious cases of self-censorship in the 20th century, nearly the entire commercial press either

ignored or significantly buried a 1936 report from a respected scientist warning that tobacco use shortened people's lives. With few exceptions, the self-imposed ban on reporting about the dangers of smoking continued throughout most of the 1940s, '50s, and '60s, a period during which major news organizations readily provided ample reporting on other health crises such as the spread of the diseases polio and tuberculosis.[13] The carriers of those diseases did not advertise, however, and thus were not treated as deferentially as the tobacco industry, which poured billions of dollars into the coffers of media companies during the 20th century.

Ben Bagdikian, a former Pulitzer Prize-winning journalist turned educator and media critic, has compared modern self-censorship with the more overt types of direct censorship practiced by advertisers and Gilded Age press barons:

> Most owners and editors no longer brutalize the news with the heavy hand dramatized in movies like Citizen Kane or The Front Page. More common is something more subtle, more professionally respectable and more effective: the power to treat some unliked subjects accurately but briefly, and to treat subjects favorable to the corporate ethic frequently and in depth.[14]

Most often this type of censorship goes undetected unless brave whistle-blowers inside the industry speak up or if diligent researchers and critics investigate the financial dealings of media companies and chart their coverage of major issues involving sponsors' or owners' interests. Thanks to the efforts of such people, there have been numerous studies conducted over recent decades, which suggest that advertiser pressure on newsrooms and self-censorship have only become worse in recent decades. It would be impossible to describe in detail the findings of these numerous studies, but we can point to some of the alarming highlights from several of the most recent ones conducted between 1997 and 2002:

- In 1997, Lawrence Soley, a communications professor at Marquette University, surveyed broadcast journalists who were members of IRE, an organization for both print and broadcast investigative journalists. Soley asked TV reporters and editors about advertiser muscling of their news operations and their stations' responses to these pressures. "Nearly three-quarters of the respondents reported that advertisers had 'tried to influence the content' of news at their stations," according to Soley's survey. In addition, "The majority of respondents also reported that advertisers had attempted to kill stories" or punish stations after a critical report aired by withdrawing advertising. Singled out for their censorial efforts were automobile dealers and the restaurant industry, both of which are major sponsors of TV newscasts.

As to whether stations yielded to advertiser pressure, Soley found that nearly as many reporters (40 percent) said their stations caved in to sponsors' demands as had resisted the pressure (43 percent). Soley also asked whether or not pressure to avoid stories came from within stations and found that over half (59 percent) of the respondents answered yes. More than half (56 percent) also said there was pressure inside their stations to produce stories designed to please advertisers—so-called "puff" pieces. As Soley reported in the magazine *Extra!*, pressure was often very subtle, consisting of suggestions from management to investigate issues not connected to advertisers or to give managers, and hence the sales staff, a "heads up" if a story might have the potential to reflect negatively on a sponsor. Reporters hoping to advance their careers are keenly aware of who does the hiring, firing and promoting in their industry and as such, they often find it difficult to disregard, or question, managers' suggestions.[15]

- In 2000, the Pew Center for the People and the Press, in association with the *Columbia Journalism Review*, surveyed over 200 local and national journalists from both print and broadcast media and found that 26 percent of the respondents admitted that they had "purposely avoided" important stories, while almost as many reporters said they'd "softened" stories to benefit the interests of their employers. In addition, 41 percent of the reporters surveyed admitted that they have engaged in "either or both" of these practices. About half of the respondents said that "peer pressure" (fear of embarrassment or potential career damage) is a factor in self-censorship, and over half say they get "signals from their bosses" to avoid certain stories.

The Pew Center also surveyed members of IRE and found that they were more likely than either local or national journalists to cite the impact of business pressures on editorial decisions. "A strong majority (61 percent) of this group," the Pew Center's director reported, "believes that corporate owners exert at least a fair amount of influence on decisions about which stories to cover; 51 percent of local journalists and just 30 percent of national journalists agree."[16]

- In both 2000 and 2001, *Extra!* reported numerous cases of advertising muscling and self-censorship in both print and broadcast media, including some of the most respected news outlets in the United States. One case in particular consisted of a special deal the *Washington Post*, the *New York Times* and the *Wall Street Journal* made with United Airlines and US Airways. The two airlines promised to give the three papers an exclusive story about their planned merger if they agreed not to print any critical comments from rivals or industry watchers. The three newspapers agreed; however, the deal fell apart after the *Financial Times*, a British-based business news outlet, broke the story on its Web site. Though the newspapers were criticized in journalism circles for making the deal, managers accepted the fact that such arrangements had become the price for being big players in business news.[17]

- Since 1998, the Project for Excellence in Journalism (PEJ), a think tank studying media performance, has released yearly reports on the quality of television news. In its 2001 edition, PEJ reported that more than half (53 percent) of the news directors surveyed said sponsors "pressure them to kill negative stories or run positive ones." One respondent told surveyors, "Sales [departments are] getting more and more influence on newscasts." Others said they were able to withstand the pressure with the assistance of their general managers. Overall, though, the study concluded, "the relentless push by advertisers and sales departments inevitably yields small concessions from beleaguered news directors. Even without overt pressure, news directors may feel obliged to compromise just to keep their jobs."

  In addition to pressure from sponsors to censor news, many news directors admitted that "advertisers get something more than just commercial time for their money." This included mention of the sponsor's name in newscasts by news announcers, appearance of the sponsor's logo in newscasts, and interviews with sponsors in the body of a newscast.[18]

  In its 2002 edition, the PEJ reported that "Not even the [September 11, 2001] attack on America and the war on terrorism could wrench local TV news" from its formulaic, advertiser-driven coverage of "carnage, crime and accidents." In addition, regular investigative, "watchdog" reporting continued to decline nationwide, likely due to pressure from sponsors and upper management as well as a lack of resources.[19]

Advertising and Target Audiences

Aside from direct censorship and the more subtle forms of pressure that result in self-censorship, advertisers influence news content in a third way that easily slips under the radar of many watchful journalists, educators and critics.

In describing the economics of commercial media, for example, a noted college textbook devoted to journalism ethics states, "while newspaper companies sell primarily one product (the paper), broadcasters have two kinds of products: news and entertainment programming."[20] Though true on a superficial level, this statement is a bit misleading. Newspapers and broadcasters do produce and sell content (news and entertainment), but content is not their *primary* product—that is, the product they actually make money from. Rather, most publishers and broadcasters earn the great majority of their money, and derive their profits, from advertisers who pay for access to audiences. Put another way, commercial news outlets sell audiences—or access to audiences— to advertisers for large sums of money. Content is clearly important, but its main purpose is to attract the eyeballs, and ears, of audience members. Furthermore, most advertisers, especially those who sponsor the news, want to attract the eyeballs of fairly privileged audience members —those who have the financial means and are likely to be frequent shoppers. Thus, if a newspaper or TV station wants to attract and retain big spending advertisers, it had better pitch its news at a carefully selected "target" audience. The broad target audience for mainstream, commercial TV news is 18- to 49-year-olds with incomes above the median and suburban or urban residences. An industry trade magazine, for example, reported in late 2001 that CNN was "delivering" 180,000 viewers between the ages of 18 and 34 to advertisers each day.[21] Newspaper target audiences are roughly the same, though they may skew a bit older. Targets for online news will also tend to be the same, with exceptions for Web-only sites aimed at niche audiences in various age and income brackets.

Demographics

An easy way to find out more precisely which audience groups, or demographics, news outlets are targeting is by reading the business and advertising press or media trade journals such as *Electronic Media*,

Broadcasting and Cable and *Editor & Publisher*. These are the publications through which media owners, managers and advertisers "talk" to each other about their businesses. An even easier method is to visit company Web sites, which usually have pages specifically directed at advertisers. Here, one can often find the type of data sales people present to current or potential advertisers. On *The New York Times* Web site, for example, an "Online Media Kit" contains nifty animations as well as several pages with profiles of the typical *Times* reader. Thanks to sophisticated data-gathering techniques provided by marketing firms and ratings services, a media company can break down its audience based on characteristics such as age, address, occupation, income, marital status, education, shopping habits, and so forth. Daily readers of *The New York Times* print edition have a median household income of over $92,000 compared to $51,000 for all adults in the New York City market and $50,000 for the United States. The average income of the *Times'* online audience is $85,000, thus both the print and Web versions of *The New York Times* attract the eyeballs of a privileged class of readers highly coveted by major advertisers.

In addition to income, the *Times'* Web site describes the favorite leisure activities of its online readers: 60 percent enjoy reading, while 21 percent enjoy gardening, 30 percent have attended a live concert in the last 30 days, and a staggering 85 percent made an online purchase in the last six months. This information is all very helpful to advertisers, especially those connected to e-commerce services wanting to sell luxury items, books, garden supplies and concert tickets. In summing up the characteristics of its audience, the *Times* boasts, "NYTimes.com readers are affluent, educated members of their communities who have a wide range of hobbies and interests. When compared to other news sites, NYTimes.com reaches more readers who represent this highly desirable target audience."[22]

The pressure to harvest target audiences for advertisers leads to another form of news filtering that is even subtler than self-censorship. Noted legal scholar and media critic C. Edwin Baker once described the primary way in which the invisible hand of the advertising market shapes editorial content by recalling an anecdote involving former *Los Angeles Times* publisher Otis Chandler. When asked why the *Los Angeles Times* did not report more on minority issues, Chandler responded that it would not make sense financially. Simply put, more coverage of minority issues would cause members of the minority community to purchase the paper, but the company loses money on each sale of a copy of the paper unless it can sell the audience member to advertisers. The *Times'* advertisers weren't interested in "buying" large numbers of minority readers because their audience research showed that, as a group, they lacked purchasing power.[23] Thus, a prominent publisher admitted that his paper deliberately avoided a large amount of news that affected major segments of the population simply because the people making up those segments weren't rich enough.

According to Bagdikian, Chandler isn't alone, not by a long shot. In his book *The Media Monopoly*, Bagdikian presents numerous examples of publishers and broadcasters altering the content of the news to make it more palatable to affluent audiences. "The standard cure," Bagdikian writes, "for 'bad demographics' in newspapers, magazines, radio, and television is simple: Change the content. Fill the publication or the programs with material that will attract the kind of people the advertisers want."[24] This is a conclusion that was also reached by John McManus, a communications professor and author of the book *Market-Driven Journalism*. As McManus explains, the rush to attract privileged audiences contradicts the conventional belief that news content simply reflects the demands of the public—that media "give the people what they want."

> Market journalism values the attention of the wealthy and young over the poor and old because news selection must satisfy advertisers' preferences. In fact, rational market journalism must serve the market for investors, advertisers, and powerful sources before—and often at the expense of—the public market for readers and viewers. To think of it as truly reader or viewer-driven is naïve.[25]

Serving Buyers and Sellers

Increasingly, Bagdikian argues, media owners respond to the advertising market by expanding the "softer" sections of their offerings at the expense of hard-hitting investigative reporting and substantive political discussion. The advent of newspaper sections called "Style," "Living Arts," "Travel and Leisure" and "Wheels" are a testament to the power of advertising. In addition to pulling resources away from serious journalism, these sections, which also fill commercial news Web sites, provide a better frame in which to view ads—the all-important *context* discussed previously. Feature stories about lush vacation spots and the clothing styles of celebrities don't usually fire the reader's critical thinking powers and thus are more likely to put them in the "buying mood."

In addition to adding "softer" news, Bagdikian, McManus and other critics contend that market journalism tends to avoid sustained investigative reporting about corporate power and the economic inequalities inherent in American-style capitalism. To be sure, the business press, TV networks and mainstream Web sites will report on the misdeeds of major corporate criminals, but they tend to give far more attention to street crime—the crimes of the poor—when reliable research indicates that corporate crime is increasing while street crime has actually decreased in many areas and corporate crime is far more threatening and costly to taxpayers than street crime. A 1999 study by the National White Collar Crime Center found that one in three American households are the victims of corporate crime.[26] Furthermore, critics argue, news reports about corporate crime tend to focus solely on the individual perpetrators—the "bad apples" in the corporate barrel—rather than the deeper issue of corporate power and its affect on democracy. As Robert Jensen, a journalism professor at the University of Texas, writes: "Business reporters are not critics of the system; at best, they police its boundaries. They report on violations of the rules; they don't ask questions about the fundamental justice of the rules."[27] Bagdikian makes a similar argument:

> Some reporters often criticize specific corporate acts, to the rage of corporate leaders. But the taboo against criticism of the system of contemporary enterprise is, in its unspoken way, almost as complete within mainstream journalism and broadcast programming in the United States as criticism of communism was explicitly forbidden in the Soviet Union.[28]

In sum, journalism funded primarily by advertising will tend to produce a primary product (audiences) that satisfies the needs of the buyers (advertisers) and sellers (media companies) in the marketplace. Similarly, the secondary product (content) will, with exceptions here and there, also tend to serve the interests of the buyers and sellers in the marketplace. This is no conspiracy; rather it is simply the way markets work, whether they be media markets or the markets for expensive consumer goods. So long as advertisers are the media's primary customers, they will have a strong, though often invisible, influence on media content.

Advertising and Corporate Structure

The censorship studies discussed earlier suggest that the much-heralded "wall" that prevents advertisers and owners from directly influencing the editorial process has always been tenuous at best and in many cases invisible. The data might even suggest that ethics codes have not been of much use in protecting the editorial process. Perhaps this is because most ethics codes are aimed primarily at front-line employees who have little power in modern corporate newsrooms. For example, the RT-NDA code mentioned earlier exhorts journalists to "Determine news content solely through editorial judgment and not as the result of outside influence" and "Refuse to allow the interests of ownership or management to influence news judgment and content inappropriately." Given the sources of pressure listed previously, maybe the codes should address sales personnel, upper management and owners and tell them it is unethical to interfere with the editorial process. To be sure, ethics are vitally important for all reporters, but no matter how hard a front-line worker tries to live up to well-meaning protocols, he or she still faces pressure to conform to the corporate culture that pervades most newsrooms today. In this environment, orders flow from the top down, not the bottom up. Unless there is collective power among rank and file journalists coupled with solidarity from principled editors and news directors, it will be difficult to stop advertisers and top corporate bosses from meddling with the news.

But even if ethics codes were pitched at corporate managers, they too would face a significant obstacle in trying to adhere to them. As Jim Naureckas, a media critic and publisher of the magazine *Extra!* points out, the legal structure of a for-profit media corporation is far different from that of an independent, family owned news outlet, thus

> A corporate owner is fundamentally different from an individual or family owner. An individually owned news outlet may be good or bad, but if the owners want to, they can pursue quality journalism even if such decisions hurt their bottom line. The management of a corporate news outlet does not have that luxury. By law, management must not allow other considerations (like journalistic ethics or the public interest) to stand in the way of corporate profits—otherwise it would be abandoning its fiduciary responsibility to its stockholders and would be subject to a lawsuit.[29]

For most of the 1980s and '90s, corporate media owners and executives were used to seeing profit margins of between 20 and 40 percent, with some broadcasters reporting margins as high as 50 to 60 percent. To put that into perspective, consider that most business sectors are lucky to have profit margins of 15 percent during their biggest boom cycles. Though the first years of the 21st century have been marked by a prolonged economic slump, most major media companies and their news divisions remain highly profitable. If anything, the pressure to improve profit margins, especially in the short term, is even greater today and as such, owners and top managers appear to be more willing to interfere directly with the editorial process or restructure their operations so that advertising departments participate in the management of news departments.

A notable example of such a reorganization came in the late 1990s when Mark Willes, an executive from the General Mills cereal company with no journalism experience, was hired to run the highly respected newspaper the *Los Angeles Times*. Soon after taking over, Willes announced that he intended to remove whatever was left of the ethical wall between editorial and advertising departments with a "bazooka" if necessary.[30] According to a former *L.A. Times* business editor turned consultant, "Willes' weapon was to create a cadre of 'general managers,' patterned after the product managers in consumer product companies like Procter & Gamble, who would create innovative marketing plans and bring editorial and advertising together to generate new revenue streams."[31]

Though the paper's circulation didn't increase as much as Willes promised, the stock price of the paper's parent company, Times-Mirror, climbed significantly. Willes was a hit on Wall Street.

Things turned a bit sour for Willes, however, in late 1999 when it was learned that on October 10, 1999, the *Los Angeles Times* devoted its entire Sunday magazine section to coverage of the Staples Center, a new sports and entertainment arena in L.A. that was about to open. In addition to preparing the special 168-page section, which consisted primarily of flattering feature stories, photos and graphics of the new arena, the paper agreed to split the $2 million in revenues from that issue with the owners of the Staples Center. The paper had, in effect, made the Staples Center co-producer of a major editorial section of its Sunday paper, without disclosing the arrangement to its readers and most of its own staff, an act seen by many as unprecedented in modern journalism.

Though the Staples Center deal and Willes' overall management approach were widely criticized by many journalists, dozens of newspaper publishers did not think the deal constituted any major ethical violation. As the *Columbia Journalism Review* reported: "After the

Times's problem came to light, *Editor & Publisher* magazine surveyed 165 editors and publishers and found that 51 percent of the publishers thought the *Times*/Staples deal was acceptable. Only 19 percent of editors agreed."[32] The views of publishers, who happen to be in the position to hire and fire editors, appear to support Bagdikian's belief that Willes "had changed the public ethics of newspapering for a wide range of his fellow publishers . . . The leader of the *Los Angeles Times* had let the moneylenders into the temple and had proclaimed loudly and proudly that this is where they belong."[33]

For those publishers and news managers who oppose Willes' way of doing business and who are genuinely concerned about the growing power of advertisers and the simultaneous deterioration of journalism, there certainly are alternatives that would truly represent "outside the box" thinking. For example, media outlets could try to become less dependent on advertising for revenue. This would likely mean giving up huge profit margins, which would not sit well with Wall Street but might make local newspapers and broadcasters more accountable to their communities. Owners and managers could also not stand in the way of efforts of journalists to organize or strengthen unions, which could serve as a countervailing power against pushy advertisers; rather than a few principled journalists standing alone, there would be strength in numbers. Concerned publishers and journalists could also try to establish nonprofit newspapers to be operated as public trusts. Revenue surpluses derived from a combination of advertising, subscriptions and direct sales could be pumped back into the paper to hire experienced investigative reporters, raise the notoriously low salaries of entry-level reporters and editors, and create media and democracy education programs. The highly respected *St. Petersburg Times* in Florida has done something like this, using a share of its profits to operate the Poynter Institute for Media Studies, which is named for the paper's former publisher Nelson Poynter, who kept his paper from being purchased by a corporate chain. Other family owned newspapers have made similar arrangements, and on the Web there are numerous nonprofit sites offering news and commentary that are funded by foundation grants, subscriptions, and donations from readers, or some combination thereof. However, these sites don't have big, if any, promotional budgets and are not featured on major portals. As such, they are not as well known as sites connected to major media companies. Typically, they tend to reach smaller audiences of experienced Web users that actively seek out alternative, noncommercial news.

On the whole, most industry leaders and journalists have accepted the corporate-run, ad-driven model as the only viable way to produce and distribute news to the public. There are a few exceptions to the advertising model in online journalism, which has been in an economic slump compared to major print and broadcast media. Some news sites are adding revenue from the sale of personal information gathered from Web users, sometimes without their knowledge. Of course this brings on a whole host of ethical problems concerning individual privacy (refer back to chapter 4). This being the case, and considering the fact that Web sites of leading print and broadcast news outlets, which are almost wholly supported by advertising, dominate Internet news ratings (see afterword on media consolidation), it could be concluded that online journalists face the same pressure from sponsors and owners to bend ethical principles that their offline counterparts do. Some have even suggested that the pressure is actually greater online due to the uniqueness of digital media platforms such as the Web. The Web might just be the best medium yet for mixing news and advertising due to its ability to make seamless transitions "between news and the products and services that lie behind the story," as one journalist has described it.[34] Thus, the economics of commercial media coupled with the specific architecture of digital technology appear to have tempted some advertisers and media owners to dissolve what's left of the fragile line between the presentation of ads and the presentation of editorial content.

Ethical Dilemmas in Online Advertising

It has been difficult for news publishers to agree on industrywide standards for the design, placement and labeling of ads on the Web, largely due to conflicts among journalists who, despite previous ethical training, can't seem to agree where the line between news and advertising is, or should be, in cyberspace.[35] This may be related to the flexible, technological structure of the Web, which is still evolving, and as such, it naturally invites experimentation. By contrast, print publishers have long been constrained by the spatial limits of the physical page (and having only so many pages per issue to work with) and broadcasters have been constrained by the temporal limits of program schedules (and being able to run only one program, or commercial, at a time). Unlike their offline counterparts, Web publishers have virtually unlimited space for content and may run multiple news items and ads on the same page 24 hours a day, 7 days a week.

The Web is also interactive in that users can click on a story link or advertisement to activate it. Thus, a Web site can instantly send a user directly from a page containing news content to a sponsor's page. A news site may also invite the user to make an online purchase directly from a news page. In the late 1990s, *The New York Times* placed a "to purchase online" button at the bottom of each book review on its Web site. The link was actually a form of paid advertising, which transported users to the Web site of Barnes and Noble, a *Times* sponsor. In addition to being paid to feature the link, the *Times* also received a transaction fee each time a reader clicked on the link. While the link was labeled, the *Times* did not disclose the transaction fees. The deal drew criticism from journalists and users who suggested that the *Times* had too closely incorporated e-commerce into its news content. Others wondered if the *Times* was being pressured by Barnes and Noble to review only books that were being heavily promoted and likely to be best sellers. In 2001, the deal between the *Times* and Barnes and Noble ended, but executives with the paper's Web site suggested they would consider similar deals in the future.[36]

It is, of course, impossible to provide newspaper readers with instant shopping opportunities, and until interactive digital television becomes standard, it is also impossible to do so with a TV commercial. In addition to the questionable practice of using shopping links, Web publishers can employ a wide range of other digital ad styles quite unlike anything in print and broadcast media. Thus, the medium can be more easily tailored to meet a particular sponsor's requests, which might enhance revenues while also raising new ethical concerns.

Popular Ad Formats

Here are some of the most widely used ad formats. Ethical concerns are discussed in the text accompanying the images.

- *Banner ads* that link to a sponsor's product site. Often these are placed in a fixed location on a Web page, which is separated from news articles. However, some banners appear in between paragraphs of a news story (see Figures 9.1 and 9.2).

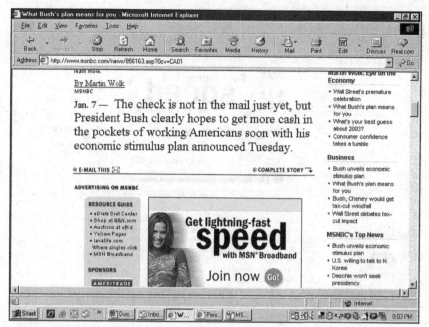

Figure 9.1 MSNBC.com banner ad inserted between paragraphs of news story.

- *Sponsored text links,* which may be placed among links to news stories (see Figure 9.3).
- *Pop-up* or *pop-under windows* that link to the sponsor.
- *Full-screen "splash" animation ads* (see Figure 9.4). As with pop-up and pop-under ads, animations are considered intrusive by many Web users. While the Salon.com animation occupies the entire page, some sites use animated "floating ads" that move across editorial content, blocking the user's view of an article, photograph or other news item. In one case, an animated ad on *The New York Times* Web site for a financial services firm inserted a bright orange animated ad with a dog wagging its tail between the second and third paragraphs of a tragic front page story of a bombing incident in Israel.[37]

In addition to these commonly used formats, some Web sites and ad agencies have developed formats that critics cite as giving the impression that the sponsor operates the page. One such format is the "wallpaper" ad used by CBSMarketWatch.com on its page containing stock

Figure 9.2 The reader has to scroll down past the ad to read the rest of the story.

quotes to promote Budweiser beer. The ad consists of numerous Bud-weiser logos built into the background design of the page—market data and headlines appear to be superimposed over the background. There is no disclaimer or explanation of the Budweiser sponsorship on the page. Web site executives say the ad is similar to a sponsorship message on an athletic scoreboard and contend that there's no journalism involved in the stock information; therefore, the ad raises no conflicts. Critics argue that the stock data falls under the umbrella of editorial content and should not be published on top of an ad.

Sponsored Content

Questions have also been raised about the presence of "sponsored content" on news Web sites. In addition to the sponsored links discussed previously, this format may consist of sponsored stories, special reports, chat features, product reviews and even entire pages or sections of a Web site paid for by a single sponsor. Such arrangements are becoming

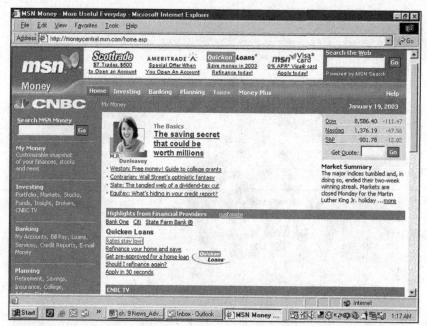

Figure 9.3 MSN's Money page includes a section called "Highlights from Financial Providers." While the links look like links to editorial content, they are actually paid advertisements that connect users to a variety of banking and investment sites. There are no labels identifying the links as advertisements, which many see as a violation of the basic ethical principle recommending that all ads should be clearly labeled.

increasingly popular on news sites primarily because they are likely to produce much higher revenues than conventional banner ads. One of the most notable examples of sponsored content that raised some ethical questions was *The New York Times'* "Tolkien Archives," which promoted the release of the movie "The Lord of the Rings: The Fellowship of the Ring." Many critics argued that this feature was poorly labeled and looked too much like a regular news section on the *Times'* page. The "Tolkien Archives" is discussed in detail in a case study toward the end of this chapter.

Sponsored content is popular with advertisers because it allows them to put their product messages very close to or within parts of a Web site that normally contain editorial content. Being placed in such a location on a respected news site may enhance a sponsor's reputation. Advertisers naturally want their brands associated with reputable content and

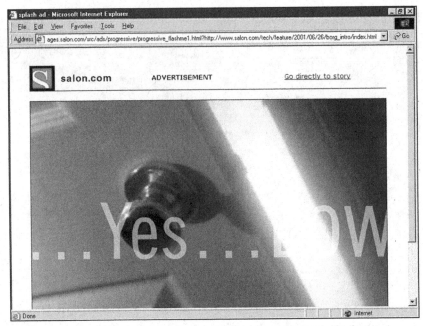

Figure 9.4 This is an advertising animation that is activated when selecting the URL of a particular article on Salon.com. The animation lasts 10 seconds, after which the article selected appears. Users who wish to see the article before the animation is over can do so by clicking a small link in the upper right hand corner of the screen.

many are willing to pay large sums of money to insinuate themselves into features that look like a hybrid of news and advertising, or that simply look like editorial features. News organizations, however, run the risk of confusing readers if they are not clear about distinguishing which pages are produced by their own reporters and which serve as marketing vehicles for sponsors.

The surfing habits of Web users also play into the popularity of sponsored content. Many news users will bypass or ignore obvious ads and marketing messages. But skillfully designed ad sections that look like news features or that mix useful information with advertisements are more likely to cause users to linger on a page longer and may induce them to access links they might have avoided if they were not so carefully integrated into editorial sections or designed to look like editorial sections.

Several journalism critics are particularly disturbed by the behavior of news organizations that allow marketing messages to be disguised or tightly embedded into news content through sponsorships. Aly Colón, a member of the ethics faculty at the Poynter Institute, argues that clear labeling and full disclosure are not enough to make sponsored features ethical. Rather, Colón suggests that news outlets should more carefully consider any arrangements that might bring their credibility into question. For example, in the situation just mentioned, users might be lured into a sponsored feature that looks like news, which might please the sponsor and raise revenue. But what is the likelihood that they'll return to the site—or visit it regularly—if after falling for the trap, they feel as though they've been deceived? If a news organization appears to be behaving sneakily and to be pursuing quick profits, Colón contends, users are more likely to lose trust, if not in the short term then very likely in the long term, which is where credibility is earned.[38]

Case Study: New Experiment in Online Advertising

For decades, long-accepted standards have governed the format and placement of paid advertisements in print and broadcast media. Newspaper ads are fairly easy to distinguish from news articles, usually offset by boxes or bold colors. When special advertising or promotional supplements (sometimes called "advertorials") are published, they usually have a different layout or design scheme from news sections. Magazines most often publish ads in this format, which typically carry the label "Special Advertising Section." In television and radio, commercials are typically clustered together in small "blocks" of time, usually two to three minutes long, and there is some separation—a second of black with no audio in the case of TV—between the program and the commercials, though this may be changing as more networks are now airing short, animated ads that float across the screen during programs. Additionally, it was reported in early 2003 that executives with the WB television network and several major advertisers are producing a variety show with no commercial breaks. Instead, commercial messages will be integrated directly into the show via the script, set design and entertainers' performances.[39] This echoes the early days of television when sponsors purchased an entire programming block and often filled it with live commercials for their products. Producers of the new show see this tactic as a way to defeat channel switching during the show and personal video recorders (PVRs) such as the popular TiVo system.

On the Web, it has been difficult for publishers to agree on clear standards for ad placement. This is largely due to the newness of the medium and its basic technological structure, which invites experimentation. For example, whereas print publishers are constrained by the spatial requirements of the physical page (and having only so many pages per issue to work with) and broadcasters are constrained by the temporal limits of program schedules (and being able to run only one program, or ad, at a time), Web publishers have virtually unlimited space for content and may run multiple news items and ads on the same page 24 hours a day, 7 days a week. Furthermore, they can adopt a wide range of digital ad styles to meet a particular sponsor's needs, including banner ads, pop-up or pop-under windows, animations, and the far more subtle sponsored link, which may be strategically placed next to, or within, editorial content.

This new publishing flexibility presents journalists with some novel ethical dilemmas. For example: How do news sites balance technological freedom and financial pressure with the journalistic imperative to keep advertising separate from editorial content? At what point does the reader become confused as to where the news ends and the ads begin? What responsibility does the publisher have to clearly label advertisements and disclose relationships with sponsors—are new standards in order? Might the placement and interactivity of certain kinds of ads be seen as inappropriate to the accompanying news articles? We address these questions in the following two case studies.

Case Number One: The Tolkien Archives

In late 2001, several weeks before it released the film "The Lord of the Rings," New Line Cinema began running an innovative movie promotion, on *The New York Times* Web site. Far greater in scope than a conventional banner ad, animation or pop-up/pop-under window, this feature was actually a "micro-site" devoted to the life and work of *Lord of the Rings* author J.R.R. Tolkien. The site, titled "The Tolkien Archives: A Guide to the World of J.R.R. Tolkien" (http://www.nytimes.com/specials/advertising/movies/tolkien/index.html), could be accessed through clickable banner ads on the *Times'* home page as well as the "Arts" and "Movies" pages. To make for a smooth transition, the *Times'* product development staff gave the Tolkien site the same look and feel as the rest of the Web site; typography, background colors and navigational features were nearly identical.

Once inside the Tolkien Archives, visitors could view a vast array of content including slideshows of artwork from Tolkien's books, audio files of Tolkien reading from *The Lord of the Rings,* an interactive "Tolkien Quiz," and a discussion forum for Tolkien fans. In addition, readers could access old *New York Times* articles about Tolkien—including his obituary and a 1954 review of *The Fellowship of the Rings* written by W.H. Auden—that had been specially selected from the newspaper's archives. Never before had the *Times* produced such an extensive online advertising package for a movie studio.

Success and criticism. In a company press release announcing the launch of the Tolkien Archives, Martin Nisenholtz, CEO of *New York Times* Digital, the company's online division, described it as a "unique, integrated marketing experience" for "online readers of the New York Times."[40] By all accounts the Tolkien site was very popular with visitors, generating over 500,000 page views in its first six weeks. By March 2002, that number had climbed to 1.5 million.

But success did not come without criticism. Several media watchers argued that the Tolkien Archives raised serious ethical questions, particularly regarding the layout of the site. Barb Palser, an online journalist and columnist for the *American Journalism Review*, commented that the Tolkien site was "virtually identical to a news section" on the *Times* Web site.[41] A study on credibility in online journalism released in early 2002 by the Online News Association (www.journalists.org/Programs/Research.htm) also questioned the look of the Tolkien site and suggested that ads that mimic "the typefaces, layout and colors used in editorial sections" deceive readers "as to the source of the information."[42] In media reports, Lincoln Millstein, an executive vice president of the *Times,* defended the look of the site and indicated that news editors had reviewed the presentation and approved it before it went online. "We try our best to control some of these things," Millstein commented. "In our five or six years of doing this [online advertising], it's a constant concern that we have with editors and others to see what is the right sensibility for the new media."[43]

Another point of contention was the label at the top of the site. Rather than use the words "Special Advertising Section," as most print publications do when dealing with special ad supplements, the *Times* elected to use the term "Sponsored Feature." Palser, along with Paul Grabowicz, director of new media at the University of California at Berkeley, suggested that the label might have been insufficient because the term was not defined nor was the sponsor directly identified. But Christine Mohan, a spokesperson for the *Times'* digital division, said the word "sponsored" is often used to identify similar content in broadcast media and is "appropriate for the Internet."[44] Nisenholtz also defended the label and the overall look of the site, though he told the *Wall Street Journal* that to avoid confusion in the future, the *Times* would use more explicit labels on similar future projects.[45]

The *Times* did not allow any current news articles about "The Lord of the Rings" movie to be used in the package, but it did publish archival material on the Tolkien site. Mohan said the *Times'* "product development and sales teams" were responsible for selecting the old articles about Tolkien and formatting them for the Web. At no time, she reports, did any members of the *Times'* editorial staff or representatives from New Line Cinema select articles.[46]

The inclusion of *New York Times* news articles within the "Sponsored Feature" was most troubling to journalist David Feld, online editor of the *Times Herald-Record*, an upstate New York newspaper. He argued that such use of news content, no matter how old, compromised the *Times'* journalistic integrity. "Anyway you cut it," Feld commented in an e-mail discussion about the Tolkien Archives, "The *Times* has turned editorial content into advertorial content."[47] Palser reported that other critics also wondered if the *Times* had compromised the integrity of its archives. "Is old news less sacrosanct than current news?" she asked in her *American Journalism Review* column.[48]

Case Number Two: A New Advertorial and a New Look

Just as Nisenholtz promised, the *Times* used a much more explicit label on its next content-based ad package. In late February 2002, "The Laramie Project Archives" (http://www.nytimes.com/ads/marketing/laramie/), a special Web site promoting an HBO movie about murdered gay college student Matthew Shepard, debuted on the *Times'* Web site. Much like the Tolkien site, it contained links to news articles from the *Times'* archives—specifically, 17 articles from 1998 to 2000 about the murder of Shepard as well as the trial of his two attackers and the debate about hate-crimes legislation that the case provoked. However, unlike the Tolkien feature, the Laramie Archives were labeled as a "Sponsored Archive" and bore the following disclaimer: "The reprinting of these articles from *The New York Times* was paid for by HBO Films, producers of the movie 'The Laramie Project.' The editorial staff of the *Times* was not involved in the selection of the articles or the production of the archive." The product development staff also distinguished the Laramie Archives from its regular news content by using a different background color and excluding common navigational features. Only a link to the *Times'* home site and "Member Center" were included.

After viewing the Laramie site, Barb Palser praised the *Times* in *American Journalism Review* for not repeating the layout and labeling "mistakes" in the Tolkien Archives. "The Laramie section," she wrote, "is at least as distinct from the Times' news pages as any magazine advertorial."[49] Mohan indicated that criticism of the Tolkien site helped the staff update its standards for similar projects. "We are pretty comfortable," she said, "with the various conventions we have adopted for sponsored features."[50]

Despite the new look and disclaimer, some critics still questioned the use of old news articles in the context of a movie promotion. Ann Brill, a communications professor at the University of Kansas, told *The Wall Street Journal* that Web features that mix news content with advertising messages "don't do readers any favors by confusing them."[51]

Sample Guidelines for Publishing Online Advertorials

According to the ONA credibility study, readers who are confused as to the source of online content are likely to become unhappy readers who may not return to the offending Web site. In hopes of protecting the boundary between news and advertising and enhancing online credibility, the American Society of Magazine Editors (ASME) adopted a set of ethical guidelines for publishing online advertisements. Titled "Best Practices for Digital Media," the ASME's guidelines specifically refer to online advertorials such as the Tolkien and Laramie Archives. Among other things, they urge publishers to make sure that

- All online pages should clearly distinguish between editorial and advertising or sponsored content. If any content comes from a source other than the editors, it should be clearly labeled . . . The site's sponsorship policies should be clearly noted, either in text accompanying the article or on a disclosure page to clarify that the sponsor had no input regarding the content.

- Special advertising or "advertorial" features should be labeled as such.
- To protect the brand, editors/producers should not permit their content to be used on an advertiser's site without an explanation of the relationship (e.g. "Reprinted with permission").[52]

In a similar fashion, the Poynter Institute for Media Studies and the American Society of Newspaper Editors developed a sample protocol for online advertising. Here are some of the highlights:

Advertising Challenges

The audience will be able to clearly distinguish between editorial content and advertising, including advertorials and other advertising models as they emerge. Advertising will not dictate news content or presentation, nor will new technologies blur the traditional distinctions between independent and paid content.

Repackaging

Repackaging of pre-existing editorial content will not compromise the integrity of the news product. All source material, including material from advertising supplements, will be fully identified as a product of the editorial department; any editing or technological enhancements that change the nature of the original product will be noted so as not to deceive the audience.

Partnerships

Relationships with partners will not compromise the basic mission or integrity of the news organization. Partnerships, such as and including site partners, content providers or sponsors, will be identified if they have a bearing on the content. News organizations that enter such partnerships will be diligent in the protection of their primary contribution, which is independent reportage.[53]

Both sets of guidelines advise readers that protocols, like the Web, are fluid and, in the words of the ASME, "subject to change as the medium evolves." Given this fluidity and the pressure to boost revenue, many publishers will continue to test or stretch old boundaries with new online advertising models. The Times' "sponsored archive" model may be displaced by a new format that further merges news elements and marketing messages. Thus each new model is likely to bring unique ethical dilemmas, which will test the journalistic and communicative values of publishers and news organizations.

Advertising Issues and the Credibility of Online News

Real and perceived conflicts of interest involving news sites and advertisers have prompted many industry insiders to take a closer look at the credibility of online news. The ONA's Digital Journalism Credibility Study, found that online news is enjoying a "credibility honeymoon." It suggests that while media users consider print and especially broadcast media to be less and less credible, they are fairly neutral about online credibility. They have a sort of "wait-and-see" attitude about the medium. At the moment, they consider the Internet to be a valuable news source and they have concerns about credibility, but, as the study's director Howard Finberg reported, their opinions aren't nearly as negative as they are about traditional media, which could be converted into positive opinions if news organizations are careful to avoid obvious conflicts of interest:

> This lack of strong opinion is where the online news media may have its greatest opportunity. It is a chance to move those neutral views to positive ground. Or, conversely, there is the real danger that the public will become critical of some practices and become as critical of online news as it is of newspapers.[54]

When asked specifically if the separation between advertising and editorial content is an important factor affecting a site's credibility, Web users overwhelmingly (95.9 percent) answered yes. But when asked to rank editorial independence as a variable affecting news credibility, news consumers listed it ninth out of 11 attributes, well below accuracy, completeness and fairness (the top three attributes) and ahead of audio/visual quality and entertainment value. The low ranking may be related to another finding that suggests that about 40 percent of Web users are confident they can discriminate between advertising and editorial content. The report also revealed that the more time people spend online, the more their confidence increases. For example, 53.1 percent of those who've visited a particular site 11 or more times said it was easy to distinguish between news and advertising.

It appears, then, that users are concerned about the integration of advertising into news content, and they view it as harmful to a site's credibility. However, their concerns are tempered by confidence in their ability to identify blurry areas and the importance they place on getting accurate, complete and fair news reports. As Finberg reports: "The findings about the separation of advertising and editorial content should be reassuring to those site managers who are trying to find new ways of attracting revenue. However, it might be too early to relax about this finding, as poorly labeled content could have a negative effect in the long run."[55]

The study briefly went beyond issues of ad placement and design and asked users whether they thought advertisers and other business interests influence how news is reported. A large majority (65.6 percent) agreed that business considerations shape the content of the news. Media workers were asked the same question and just over half (51.5 percent) agreed that business interests influenced news coverage. While media professionals are less likely to say advertisers influence the news, they do tend to be more suspicious than users about the overall credibility of online news, which led to the study's authors to pose the question we noted in the last chapter: "Is there something the media perceives or knows about the ethics and practices of online news organizations or operations that the public does not know? Or are traditional media just being resistant to online news?"[56]

On one level, it may very well be that journalists in print and broadcast media feel a bit threatened by the new medium. Or rather, the fact that the Web enables instantaneous publishing may cause journalists to believe that online content will inherently be less accurate and balanced than print or broadcast news. On another level, perhaps veteran journalists who are well aware of the power of advertisers and owners and who are technically savvy know how much easier it is to hide marketing messages in news content on the Web. Given the pressure to make quick profits, do they fear that online news organizations will be tempted to further blur the lines—or do they know of news sites already doing so—which might cause irreparable harm to their profession? Whatever the case, the ONA study raises important questions that surely deserve more study and discussion among both scholars and working journalists.

ONA Recommendations

While the ONA report stops short of recommending that publishers experiment with revenue models that de-emphasize advertising, it does recommend that publishers take extra care to clearly label all paid ads and disclose the financial arrangements behind them. Additionally, the study suggests that news organizations invest in training programs that include sessions in journalism ethics and that they establish standards and policies—to be published on their Web sites—that broadly govern both editorial and advertising decisions.

Such policies for online advertising could attempt to resolve several general concerns identified by the ONA, such as (1) advertising technology issues that impact credibility, such as pop-up and pop-under ads, and the ability to serve ads to be adjacent or not adjacent to specific types of editorial content; (2) conflict of interest issues with editorial product reviews and links that enable users to purchase the products; and (3) disclosure of relationships with news partners and advertisers and the explanation of journalistic practices, processes and beliefs of the online news organization.

In addition to the concerns highlighted by the ONA, Palser suggests that news sites answer several additional questions related to specific issues in online advertising:

- Do you have standard banner sizes and layout positions for ads?
- Do you label advertisements and sponsored content?
- Do you include advertisements in news e-mail services?
- Do you ever intersperse advertising "headlines" or hyperlinks with news headlines?
- Do you accept political advertisements? If so, do you clearly identify them as advertisements? Do you position them near election-related news content?[57]

One company that has developed and posted an editorial/advertising policy is CNET Networks, which operates the popular technology news sites ZDNet.com and CNET News (news.com; see Figure 9.5). Its policy states, "Ethical considerations—such as clearly distinguishing editorial content from commerce content—are of the utmost importance so as not to confuse or deceive users" and promises that CNET will

- disclose business relationships in editorial content whenever relevant;
- label all ads as ads;
- disclose the business model behind the ads, including referral fees CNET receives from merchants;
- not review its own services or services with which CNET has a business affiliation; and
- label content from external sources.

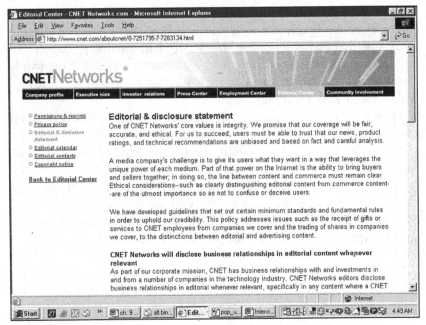

Figure 9.5 CNET Networks' "Editorial & Disclosure Statement."

In addition, CNET's policy states that "employees may not accept any gifts of value" from sponsors or other businesses that would "give the impression that CNET Networks is beholden to that company." While the company doesn't prevent employees from making investments, its policy does recommend that they "go out of their way to avoid any appearance of impropriety in regard to securities transactions or inside information about any company we cover."[58]

Other media watchers have also weighed in with recommendations about how to earn or maintain credibility. Robert Magnuson, a former business editor turned media consultant, suggests that news organizations can succeed journalistically in today's ad-driven business climate by hiring editors who have a keen understanding of the media business but who are also fiercely independent. "The editor must be a leader," Magnuson argues. "He or she must have the chops for the job, which in part means staying connected to the newsroom and not becoming overly enamored of life in the executive suite." In addition to hiring ethical editors, Magnuson believes that news organizations must also empower them to make critical decisions when potential conflicts arise. He suggests that companies allow editors to have final say on how all content is used or appears in the product, be fully informed of all significant business initiatives, and have veto power over anything that makes his or her gut uncomfortable.

Magnuson, who participated in the Poynter Institute's "Journalism and Business Values" project, also rejects the notion that journalists should refrain from learning about or being involved with business departments and vice versa.

> As unorthodox as it might feel to the editorial side, new hires on the business side need to be exposed to the newsroom and newsgathering process. They need to understand how it all comes together and how and why decisions are made. Similarly, new editorial hires should receive an intensive orientation to the business operation. They should understand the company's budget, including editorial's share. They should understand ad rates and circulation economics.[59]

As with many other industry insiders, Magnuson does not question the basic economic structure of ad-driven journalism. Instead he, like they, focuses on how best to work within the existing structure. From this perspective, he steers away from the rhetoric of the past—the "church-state wall"—and suggests an organizational model that treats both journalists and sales personnel as allies in the common project to make their news organizations credible and highly profitable.

The Ethics of "Selling Eyeballs"

Another insider who addresses ethical concerns from within the boundaries established by the industry is Merrill Brown, former editor in chief of the popular news site MSNBC.com. The ONA study mentions that he "upped the ante on standards training" during his six-year tenure at MSNBC.com, which ended in 2002. The study also quotes Brown as saying that his company "constructed a big, fat notebook," with guidelines on how to handle a wide range of ethical dilemmas including "financial disclosures for reporters and editors" (whether or not they own stock in or have personal dealings with the companies they cover). However, Brown also made it clear in an interview with the *Online Journalism Review* that the purpose of online news, just like offline news, is to serve advertisers by delivering target audiences to them. "We are in the business of maximizing eyeballs for advertisers . . . We're selling eyeballs here," as Brown emphatically put it.[60]

Critics contend that as long as journalism is primarily about connecting advertisers with target audiences in order to generate large profits, there will be conflicts between the business of "selling eyeballs" and the ethical imperative to serve the public with critical, watchdog reporting. Reporters and editors concerned about the credibility of journalism may focus their efforts on establishing standards to govern the design and placement of ads as well as encouraging their bosses to be more transparent about the business deals behind the ads. Or they might consider having a discussion about the ethics of ad-driven journalism and corporate media power in relation to the needs of democratic society. It's worth asking whether tinkering with advertising guidelines and adding disclosure notices will sufficiently resolve conflicts of interest, or are structural changes needed? If so, is it possible to work collectively to experiment with revenue models that are not so tied to advertising dollars? Is it possible to consider establishing news outlets that function independently of large corporations, which are legally bound to serve investors and shareholders first and foremost? Perhaps in the discourse about journalism ethics, there is room for these questions too.[61]

Summary

There has been a long-standing belief in journalism that the business and editorial sides of a news organization should be clearly separated in order to prevent advertising and other business pressures from influencing news content and damaging credibility. The metaphor that is often invoked to describe this ethic is that of an impenetrable "wall" between the two operations. The emergence of online news sites, with their novel display and interactivity capabilities, has created new challenges in how the wall can be maintained and even whether it needs to be maintained at all in the new media environment.

It is possible to trace the beginning of this journalistic ethic of separating advertising and editorial by looking at journalism as it evolved in the early and middle 20th century. At this time, the news business began relying more on advertising, and media ownership was becoming more consolidated. These trends gave rise to concerns about the credibility of newspapers, which led to a movement toward professionalization. Professional journalism followed an objective news reporting model, which required an effort and awareness to keep the business side of the organization away from the editorial. This "church/state" type separation became a fundamental journalistic ethic and was eventually cited as such in the earliest codes of ethics of the major media organizations.

While this separation cured some of the most obvious abuses and influences of the business side over editorial, in practice, the wall has been tenuous and sometimes even nonexistent. In fact, there have been many examples of business and advertising pressures influencing news operations over the years, whether through instances of direct censorship, self-censorship or editorial-related decisions that were made solely to please and attract advertisers.

In the online news environment, there has been conflict and confusion over whether new forms of online advertising—banner ads, sponsored text links, pop-ups/pop-unders, animation, wallpaper and sponsored content—are confused by readers as editorial or are damaging editorial integrity. According to recent studies of online news credibility, online news users are concerned about the integration of advertising into news content online, but they also believe that they have the ability to identify the blurry areas.

As a way to guard the separation between news and advertising, the American Society of Magazine Editors, the Poynter Institute for Media Studies, and the American Society of Newspaper Editors have all created recommended guidelines. Certain online news sites, such as MSNBC and CNET Networks, have also posted an explicit statement outlining their policies for editorial independence.

Critical Thinking Questions

- Do you think that the traditional "wall" between editorial and advertising is a fundamental journalistic ethic? Why or why not?
- What kinds of online advertisements described in this chapter are least likely to be confused by readers as editorial matter? Which are the most likely to cause confusion?
- Is there a way for a news site to include sponsored text links and clearly distinguish them from editorial? How?
- If you were writing an article online that mentioned a company, would you see any problem in providing a link to the home page of its Web site? Would doing this be more a reader service or a free advertisement?

- If online users eventually become accustomed to a less strict distinction between editorial and commercial material online, are online news sites still ethically obligated to maintain a strict separation? Why or why not?

Key Terms

Banner ads
Pop-up/pop-under
Sponsored content
Sponsored text link
Wall between editorial and advertising
Wallpaper

Recommended Resources

Books

Baker, C. Edwin. *Advertising and a Democratic Press* (Princeton, NJ: Princeton University Press, 1994).
Barnouw, Eric. *The Sponsor* (New York: Oxford University Press, 1979).

The Internet, Media Consolidation and Democracy

This book's focus has been on the ethical dilemmas and choices that confront the individual online media professional. And as we said in the preface, we strongly believe that by understanding the principles at stake, and by making ethical choices, individual journalists, editors and newsroom managers will move this newest journalistic medium forward in a positive direction, help create a model of quality journalism, and contribute to an overall healthier media environment.

But there are also larger issues and forces that play a behind-the-scenes role that is shaping how this medium develops. These are the legal and regulatory forces that establish the actual structure of the media: expanding or limiting the powers of the existing media firms and determining the architecture and capacities of the Internet. As a media professional you also need to be aware of these larger political, legal and regulatory activities and decide if there is a part you can also play on the larger stage and have a say in how the media world develops.

Increase in Media Concentration

Over the last few decades, large media companies have merged and expanded at an astonishing rate. Such consolidation has been worrisome to those who are concerned about the ability of the media to be an independent voice that can report on both governmental and corporate uses and abuses of power.

Former journalist Ben Bagdikian began documenting media concentration in the early 1980s. In the first edition of his book *The Media Monopoly* (1983), he sounded the alarm that the U.S. media system was dominated by just 50 companies, with enough power and influence to constitute a "Private Ministry of Information and Culture."[1] By the year 2000, Bagdikian would report that the number had dropped to six.[2] What's more, the top six companies have become some of the wealthiest companies in the U.S. economy and major powers in global media, distributing news, entertainment and advertising around the world.[3] Because they are vertically integrated, with interests in almost every media sector—newspapers, magazines, movie studios, record labels, radio, broadcast television, cable television, theme parks and professional sports franchises—these giant companies can easily reuse content and cross-promote products across platforms. This effectively eliminates any remaining competition, in the true economic sense, from U.S. mass media. To be sure, there is often bitter rivalry among the conglomerates, but economic rivalry between oligopolistic powers is not the same as genuine competition.[4] As Bagdikian puts it:

> Under the folklore of capitalism, even giant firms would compete forcefully against each other. But through mutual cooperation, interlocked directors, and shared partnerships in media operations, contemporary capitalist competition has become more like a cooperative cartel.[5]

There can be no doubt that the Internet and the World Wide Web have changed the media environment in general, and journalism in particular, in the United States and other advanced industrial countries by expanding the number of news and information outlets consumers can choose from and empowering some consumers to become media producers themselves. However, there is quite a bit of doubt as to whether the Net and the Web have significantly tilted the balance of media power in these nations away from the consolidated giants or whether digital communications technology will serve as democratizing agents in the developing world.

The doubts arise from observations about recent political, economic and cultural developments, which do not seem to confirm the utopian predictions made by several early cyberspace theorists. Those predictions, as summarized by journalist and media critic James Fallows, included claims that "the Net would simply render unimportant the networks and large media companies," that it would give poor and oppressed peoples easy access to information and two-way communication, thus "undermin[ing] tyrants," and that it would enhance existing democracies by reducing the cost of political communications (advertisements) while enabling people to become more politically informed and active.[6] In short, we were told that online communication would radically alter social relations and contribute to the "fall of the media establishment."[7]

However, many observers have long argued that such promise was unrealistic, even hugely overblown, while others have suggested that recent trends in cyberspace might even enhance the power of the media establishment and have a marginal effect on political participation. Could it be that much of the hype surrounding the Internet, in both its "information superhighway" and competitive "cybermall" incarnations, was just that—hype? Is it the case, as several communications scholars warned in the mid- and late 1990s, that the wild and competitive Internet frontier is being tamed or "normalized" by a group of vertically integrated media conglomerates, with the help of political forces in Washington?[8] Have small entrepreneurs and the nonprofit, civic sector been pushed "to the distant margins of cyberspace," from where they are not likely to threaten the consolidated media powers of the broadcast era?[9]

Behind the Hype

Before moving on to a summary of specific Internet trends, it is worth noting that utopian rhetoric is not unusual when new technologies emerge in industrial societies. It is, in fact, quite the norm. As Robert W. McChesney observes, "every major new electronic media technology this century, from film, AM radio, shortwave radio, and facsimile broadcasting to FM radio, terrestrial television broadcasting, cable TV, and satellite broadcasting has spawned similar utopian notions."[10] The development of television in particular inspired elegant promises of empowerment and social transformation. In the 1940s, a scientific journal went so far as to suggest that television would foster global understanding and cooperation. "International television," the magazine promised, "will knit together the peoples of the world in bonds of mutual respect; its possibilities are vast indeed."[11]

But while scientists and communications engineers marveled at the potential for global communication and unity, the advertising industry consistently hailed TV as a "magical marketing tool" based on its ability to transmit seductive lifestyle images and sounds into the home.[12] This, of course, reminds us that "technology evolves within our existing social and economic and political contexts" and not in a vacuum.[13] In the United States, television evolved shortly after the regulatory structure of broadcasting, which favored centralized, commercial media companies over nonprofit community broadcasters, was put in place. Thus, it also developed into an advertising-based medium, and as such, programs most likely to attract eyeballs and, hence, advertisers, became the standard fare.

It appears as though there was a similar divide between many in the technology community, who actually designed the Net and the Web, and those in the media and political establishment as to what kind of change the Internet would bring. The computer scientists tended to refer to digital communications networks as empowering and educational resources (an "information superhighway") that would democratize communication. On the other hand, commercial interests, especially those developing new media projects, saw them as the perfect platforms from which to launch personalized shopping and entertainment services, which they claimed would benefit media providers, consumers and the overall economy. Either way, technologists and media marketers agreed that the Internet would have a positive influence on society. Some scholars have suggested that this past—as well as current—enthusiasm for the Internet is connected to the privileged economic and educational status of the members of groups who have a lot to do with designing new technologies and shaping regulatory policies. "The hype about cyberspace development," according to George Gladney and James D. Ivory, "reflects almost exclusively a narrow view from the vantage of the wealthy and privileged, and comes mostly from the technology's developers and its leading-edge users."[14]

The Problem of Access

While those behind the development of the Internet and those with high-speed access may have enjoyed some of the liberating affects of the technology, most of the rest of the world hasn't yet been online. Statistics concerning Internet access and usage reject the notion that computer networks and Web sites create a global village or public sphere, which can subvert concentrated political power and make the unrealized dream of direct democracy a reality.

The first requirement of a healthy public sphere is equal access to the means of communication, according to Jurgen Habermas. Citizens, especially those outside the circle of power politics, would need to be connected and fully represented before cyberspace could ever be considered a potential site for democratic discourse. This being the case, it is hard to view the Internet as a democratizing agent when it is estimated that only about 10 percent of the world's population have access to it.[15]

Recent statistics show that there were well over 500 million users worldwide in early 2002 and 200 million Internet hosts by June 2002. But usage is highly skewed toward the industrial nations of the West, especially the United States. For example, in 2000 the United States had more than 30 percent of all computers in use worldwide, followed by Japan at 8.7 percent, Germany at 5.7 percent and the United Kingdom at 4.7 percent. Not only is there great disparity in access to communications technology worldwide, there is also inequality of access—a "digital divide"—within the United States based on household income, race and gender. Internet penetration is highest among white, middle and upper income families, with upscale white men being the most regular users. Women, African-Americans and Latinos are beginning to access and use the Net in larger numbers at work, school and public libraries, which is slowly helping to close the gap. However, scholars studying Internet access and usage argue that growing economic disparities and reductions in public infrastructure projects are likely to keep many low-income families from using the Internet on a regular basis.[16]

Digital Infrastructure

Longtime critics of Internet triumphalism have often pointed out that they began to have concerns that the Net was being falsely promoted as a democratizing agent in the early 1990s when the U.S. government's "National Information Infrastructure: Agenda for Action" was released. The document called for most of the basic Internet infrastructure, which was built with taxpayer dollars, to be transferred to the private, commercial sector. Under this plan, critics argued, the Internet would develop on a trajectory similar to that of television and radio.[17] By the mid-1990s, most of the infrastructure was privatized and the original ban on Internet commercialism was lifted against the wishes of many early developers. Net utopians often contended that the architecture of the medium was so deliberately anarchic that it could never be controlled in the same manner as broadcast media. Besides, they asserted, barriers to entry would be so low that fierce competition would prevent traditional corporations from regulating cyberspace to better suit their needs. Critics shot back that the architecture could be changed and the medium brought under tighter control if the legal and technological rules, or "codes,"[18] governing the operation of the Internet and telecommunications policy in general were changed to favor big commercial players.

A major step in that direction occurred with the passage of the Telecommunications Act of 1996. The law bundled the various parts of the Internet's communications hardware—telephone wires, cable connections and wireless networks—into one regulatory category, which government regulators claimed would pave the way for competitive markets that would keep the price of cable and phone services down. Previously, these services had been regulated as separate industries, which limited the growth of media conglomerates. The 1996 act unleashed a flurry of media mergers in the late '90s that had the effect of reducing competition and raising cable and phone rates.[19] Greater consolidation in the cable industry has been particularly troubling to consumer groups and media critics because cable wires are the favored delivery mode for high-speed "broadband" Internet connections. Telephone companies offer Internet access through DSLs but at lower download and upload speeds than the "fat" cable pipelines. Cable connections to the Net can also be packaged together with cable TV service, which would give major cable companies near monopoly control over access to two major sources of news and other cultural programming.

Cable Mergers and Open Access

Another factor worrying critics is the merging of service providers with media content owners. When AOL merged with Time Warner in 2000, the resulting conglomerate owned a massive network of cable lines and dozens of content providers across media formats (movies, music, books, magazines, and cable TV channels) and had access to millions of Internet subscribers. Mergers have continued apace into late 2002. In November, the FCC approved Comcast's $47.5 billion buyout of AT&T Broadband. The new company will control about 30 percent of the country's cable TV market with similar domination in the broadband market. In fact, it increasingly looks as though these two giant firms will effectively control access to TV and the Internet for the majority of Americans.

In March 2002, the FCC granted the wishes of cable companies and classified cable Internet modems as an "information service" rather than as a telecommunications or cable service. This exempted cable providers from "open access" requirements that have traditionally applied to Internet services operated by telephone companies. Under such rules, large providers are required to share their services with numerous small, independent services, which gives customers a wide range of choices for Internet service. With the open access principle being scrapped, consumer groups worry that the number of small ISPs offering inexpensive connections will drop significantly and consumers will be left with fewer choices.

In announcing the new classification, the FCC said the requirements would slow down the broadband deployment process.[20]However, as law professor Lawrence Lessig points out, this argument weakens when one looks just north of the U.S. border into Canada and to South Korea.[21] In both countries, open access principles have been the norm for cable broadband networks and deployment has proceeded at a much quicker pace than in the United States. The Center for Digital Democracy (CDD), an Internet consumer advocacy group, reports that a Canadian nonprofit firm, called Canarie, "has been funded in part by the Canadian federal government to promote high-speed infrastructure deployment and applications development." The CDD also reports, "Canarie does more than simply push broadband technology at all costs—its strategies are undergirded by a commitment to building communications networks and applications that serve the public interest, including both commercial AND non-commercial uses."[22]

It should be noted that, according to the Center for Public Integrity, AOL-Time Warner was the biggest media donor to U.S. political campaigns between 1993 and 2000, with a total expenditure of more than $4.6 million. AT&T-Comcast was a major donor as well, filling campaign coffers with $1.8 million over the same time period. Both firms have also funded extensive lobbying efforts aimed at convincing Congress and FCC staffers to support broad measures designed to shift regulatory power from the public to the private sector.[23] In an interesting twist, though, the cable companies have continued to vigorously support open access requirements for telephone companies providing Internet service.

Aside from changes in cable regulation, the FCC in June 2003 also raised the cap on the number of local TV stations one company can own and eliminated the rule that prevented one company from owning a daily newspaper and TV station in the same market—the so-called "cross-ownership" rule. These changes are likely to affect the Internet because both major newspapers and TV networks have established themselves as online news providers. With more outlets in print and broadcast media, large media companies will have more material available for use—or reuse—on their Web sites and will be better positioned to promote their online offerings via their traditional media, making online advertising packages more attractive to potential sponsors.

Broadband Architecture and Discriminatory Routing

With control over much of the physical components of the Internet, cable operators are rebuilding much of the system's architecture on "center-to-end" principles rather than the "end-to-end" standard that the Net's original developers preferred. To give one example, download speeds on broadband connections are many times faster than upload speeds, which tilts the power of dissemination further away from individual users. This is consistent with the plans of commercial content providers to convert the Web from a medium in which users "pull" content from Web sites to a medium in which content, including animated advertising, is "pushed" toward users without them having to interact much with a mouse or keyboard.

At another level, discriminatory routing systems and protocols designed to steer consumers to major content providers are being developed. In a sense, then, cable companies may have more control over where customers go once they connect to the Net. Traditionally, the Internet has been a neutral platform. That is, messages are treated equally as they are routed through cyberspace; traffic jams are resolved without regard to the content of a message or the addresses of the sender and receiver. However, in the absence of public interest protections, nothing can stop cable operators from changing protocols—the agreements between computers that facilitate data transfer—in order to speed up delivery of favored content or slow down the flow of content that is not produced by their subsidiaries or corporate partners. Thus, a customer wanting to connect to independent news outlets, educational Web sites or nonprofit groups might encounter longer download times and other obstacles in getting to his or her destination. Conversely, nonprofit groups, schools and independent news services may have to make expensive upgrades to their Web sites and servers before they can be bumped up to a faster lane of traffic.

The broadband industry is also planning to introduce a variety of "tiered pricing" systems for customers. Under one model, Internet subscribers would be allotted a limited amount of bandwidth to use per month. If the customer went over the limit, he or she would have to pay extra fees. A limit of 5 gigabytes per month, according to the CDD's Jeff Chester, would restrict regular use of emerging applications such as Web radio, streaming media and video on demand. Only those with greater financial resources would be able to use such services.[24] Cable providers could also use a tiered system to charge higher fees for access to content on competing networks, portals and Web sites. For example, if a user connecting to the Net through brand X's service wanted to download music, movies or even watch streaming news video from brand Y's Web sites, they might have to pay extra to do so.

The Economics of Internet News

Consumer advocates and numerous media critics contend that the cable industry's increasing power over the Internet is likely to reduce the already low likelihood that digital entrepreneurs and independent news outlets can compete with major entertainment and information services. Media critic Norman Solomon reported in 2001 that many people who connect to the Net through a major commercial portal or ISP, such as AOL, don't browse far beyond the portal or its affiliated sites.[25] This was also reflected in a 1999 study by Lada Adamic and Bernardo Huberman, which indicated that the most popular 5 percent of Web sites accounted for almost 75 percent of user volume.[26] In 2001, Huberman observed that the top 0.1 percent of Web sites attracted a staggering 32 percent of user volume. "Such a disproportionate distribution of user volume among sites," Huberman explains, "is characteristic of winner-take-all markets, wherein the top few contenders capture a significant part of the market share."[27]

When the ratings of U.S.-based news sites are reviewed, it's clear that media conglomerates with operations in traditional media are far and away the most dominant. For example, in October of 2001, the five most popular news sites according to the Jupiter Media Metrix rating service were:

1. CNN.com
2. MSNBC.com

3. Time.com
4. ABCnews.com
5. NYTimes.com[28]

It's worth noting that Time.com and CNN.com are both owned by AOL-Time Warner, the largest media company in the world. In February 2002, another Jupiter Media Metrix survey indicated that CNN.com and MSNBC.com had switched places, and that Yahoo! News jumped to the number 3 spot while ABCnews.com and the Tribune Interactive family of newspapers (*L.A. Times, Chicago Tribune*, etc.) and broadcast affiliated (WPIX-New York, WGN-Chicago, KTLA-Los Angeles) Web sites rounded out the top five.

The emergence of Yahoo! News might appear to be proof that online insurgents can break into the ranks of the media elite. However, the popular Web portal doesn't produce any original content; rather it aggregates news content from major wire services and Web sites, including material from sites that dominate the ratings. And according to the *Online Journalism Review*, "Many of the top sites are not even attempting to cover breaking news. With few exceptions, the race among the most-read online news sites has turned into a competition to see which site can post wire copy the fastest."[29]

McChesney argues that these developments should come as no surprise. He points out that "long before the dot com frenzy turned south . . . venture capitalists stopped bankrolling Web media content ventures." Investors pulled out, according to McChesney, because they realized that "the Internet is not sufficient to introduce competition to commercial media markets." This is primarily due to the huge advantage that traditional media powers have over upstart competitors. As McChesney explains,

> They have digital programming already and need not create something entirely new; they have known brand names; they can advertise their Web activities in their traditional media, a striking advantage over someone who has to pay for that level of promotion; they can bring their advertisers over to their Web sites; they get pole position from browsers, portals, and search engines. Plus, the media giants have a much longer time horizon with regard to the Web. They will take huge losses every year because they know that by doing so they can make sure no one can use the Internet to destroy their existing empires.[30]

Thus, while it is certainly true that there are millions of Web sites offering news and opinion, many of which might break stories and cultivate lively discourse among niche constituencies, "the journalism that receives funding from advertisers, and attention, will do so under the same commercial auspices and terms that offline journalism operates under."[31]

Finally, though, we don't want to paint an overly pessimistic picture regarding the potential of the Internet as a source of alternative news. While these larger forces are in motion, there are positive signs occurring too. Although it is true that the traditional and largest media firms are the ones that are currently capturing the most online readers, there's been a great surge in smaller special interest sites, as documented by the Pew studies. And there's no denying that for the first time, users can now instantly, easily and freely get news stories with perspectives from around the globe—as easily as a click on the Google News page.

Also encouraging is the expansion in the number—and perhaps even influence—of the Weblogs. Cynthia Webb, technology writer for *The Washington Post*, called the year 2002 "The Year of the Blog," as the number of these most intriguing forms of first-person accounts swelled to the hundreds of thousands. What about their influence? It's early to say, but their impact appears to be increasing. In December 2002, commentators widely credited the Weblog community as the force that kept alive the story of Senate Majority Leader Trent Lott's comments at the Strom Thurmond 100th birthday party by surfacing earlier similar remarks by Lott and by bringing the story to the attention of the mainstream media. In fact, *New York Times* columnist Paul Krugman even cited one blogger, "Atrios" by name (http://atrios.blogspot.com) as "play[ing] a key role in bringing Mr. Lott's past to light."[32]

There are, then, several forces at work that will determine the future direction of the online media and the media landscape as a whole. The ethical media professional will work on all fronts to try to ensure that the highest values of his or her profession are the ones that ultimately emerge.

Convergence Journalism

Given the present economics of the Internet and the prospect of the newspaper/TV cross-ownership rule being dismantled or abolished, it is possible that the type of journalism that receives the most attention and funding will be "convergence" journalism. As the name implies, convergence refers to the merging of equipment and personnel from various media formats (print, TV and the Web) and news outlets into centralized news production, distribution and promotional operations. A classic example of convergence can be found in Tampa, Fla., where the journalists from a major daily newspaper, a TV station and a Web site all work in the same newsroom.[33] Convergence is meant to increase revenue while keeping production costs down. It all works if (1) workers produce a steady stream of entertaining, multimedia content that can be reused across media formats and consistently attracts target audiences (upscale 18- to 49-year-olds); and (2) sponsors with deep pockets are convinced to buy access to those audiences through expensive, multimedia advertising packages.

Media executives and consultants promise that well-managed convergence will not only help boost profit margins but improve journalism as well. It seems pretty clear that a tightly run, multimedia news combine can do the former, but what about the latter? That depends on what one considers good journalism. If detailed investigative reporting and sustained critical coverage of public affairs is good journalism, one might end up disappointed with convergence, which tends to value speed, quantity and content adaptability above all else. Stories about major spectacles and events (the Super Bowl, the Olympics, natural disasters, etc.) with familiar character types or well-known personalities are more easily inserted into the news flow of another medium.

Reporters working a convergence environment that is not well staffed may find themselves under constant pressure to churn out stories for two or three media formats each day. In some convergence journalism environments, "backpack journalists" as they are called are required to handle the technical aspects—photography, sound recording, Web layout and graphic design—of each story they've been assigned. Reporters working under these conditions are likely to spend less time gathering critical background information and conducting interviews with diverse sources.

It's worth wondering how this trend might affect the independence and abilities of the reporter to do his or her job in a quality manner. For instance, might the lack of time mean less original reporting and a necessary increase in the use of official spokespeople and data supplied by the public relations industry? And could it mean that a reporter will feel more cautious about criticizing or exposing the ethical conflicts of a media partner? In effect, the ability of the media to police itself through internal criticism—which is already tenuous—may deteriorate further at a time when reflection, discussion and review are needed most.

Recommended Resources

Books

McChesney, Robert W. *Rich Media, Poor Democracy: Communications Politics in Dubious Times* (Champaign: University of Illinois Press, 1999). Paperback edition by New Press, New York City, 2000.

McManus, John H. *Market-Driven Journalism* (Thousand Oaks, CA: Sage, 1994).

Schiller, Herbert I. *Information Inequality: The Deepening Social Crisis in America* (New York: Routledge, 1996).

West, Darrel. *The Rise and Fall of the Media Establishment* (Boston: Bedford/St. Martins, 2001).

Organizations and Institutes

Center for Digital Democracy. Web site: www.democraticmedia.org.

Columbia Journalism Review's "Who Owns What." Web site: www.cjr.org/owners/.

Federal Communications Commission. Web site: www.fcc.gov.

Media Access Project. Web site: www.mediaaccess.org/.

Media Channel. Web site: www.mediachannel.org/ownership/front.shtml.

Media Education Foundation. Web site: http://www.mediaed.org/.

Project for Excellence in Journalism. Web site: http://www.journalism.org.

Appendix

Ethical codes of major online media organizations included in the appendix:

- American Society of Newspaper Editors (ASNE)
- Associated Press Managing Editors (APME)
- International Association of Business Communicators (IABC)
- National Press Photographers Association (NPPA)
- Radio-Television News Directors Association (RTNDA)
- Society of American Business Editors and Writers (SABEW)

Following the codes of these organizations, we have also reprinted the ethical code of Gannett Newspaper Division. (Note: The code of ethics from the Society of Professional Journalists [SPJ] is reprinted in Chapter 1.)

ASNE Statement of Principles

(Published August 20, 1996; Last Updated August 28, 2002)

ASNE's Statement of Principles was originally adopted in 1922 as the "Canons of Journalism." The document was revised and renamed "Statement of Principles" in 1975.

PREAMBLE. The First Amendment, protecting freedom of expression from abridgment by any law, guarantees to the people through their press a constitutional right, and thereby places on newspaper people a particular responsibility. Thus journalism demands of its practitioners not only industry and knowledge but also the pursuit of a standard of integrity proportionate to the journalist's singular obligation. To this end the American

Society of Newspaper Editors sets forth this Statement of Principles as a standard encouraging the highest ethical and professional performance.

ARTICLE I—Responsibility. The primary purpose of gathering and distributing news and opinion is to serve the general welfare by informing the people and enabling them to make judgments on the issues of the time. Newspapermen and women who abuse the power of their professional role for selfish motives or unworthy purposes are faithless to that public trust. The American press was made free not just to inform or just to serve as a forum for debate but also to bring an independent scrutiny to bear on the forces of power in the society, including the conduct of official power at all levels of government.

ARTICLE II—Freedom of the Press. Freedom of the press belongs to the people. It must be defended against encroachment or assault from any quarter, public or private. Journalists must be constantly alert to see that the public's business is conducted in public. They must be vigilant against all who would exploit the press for selfish purposes.

ARTICLE III—Independence. Journalists must avoid impropriety and the appearance of impropriety as well as any conflict of interest or the appearance of conflict. They should neither accept anything nor pursue any activity that might compromise or seem to compromise their integrity.

ARTICLE IV—Truth and Accuracy. Good faith with the reader is the foundation of good journalism. Every effort must be made to assure that the news content is accurate, free from bias and in context, and that all sides are presented fairly. Editorials, analytical articles and commentary should be held to the same standards of accuracy with respect to facts as news reports. Significant errors of fact, as well as errors of omission, should be corrected promptly and prominently.

ARTICLE V—Impartiality. To be impartial does not require the press to be unquestioning or to refrain from editorial expression. Sound practice, however, demands a clear distinction for the reader between news reports and opinion. Articles that contain opinion or personal interpretation should be clearly identified.

ARTICLE VI—Fair Play. Journalists should respect the rights of people involved in the news, observe the common standards of decency and stand accountable to the public for the fairness and accuracy of their news reports. Persons publicly accused should be given the earliest opportunity to respond. Pledges of confidentiality to news sources must be honored at all costs, and therefore should not be given lightly. Unless there is clear and pressing need to maintain confidences, sources of information should be identified.

These principles are intended to preserve, protect and strengthen the bond of trust and respect between American journalists and the American people, a bond that is essential to sustain the grant of freedom entrusted to both by the nation's founders.

APME Code of Ethics

(Revised and Adopted 1995)

These principles are a model against which news and editorial staff members can measure their performance. They have been formulated in the belief that newspapers and the people who produce them should adhere to the highest standards of ethical and professional conduct.

The public's right to know about matters of importance is paramount. The newspaper has a special responsibility as surrogate of its readers to be a vigilant watchdog of their legitimate public interests.

No statement of principles can prescribe decisions governing every situation. Common sense and good judgment are required in applying ethical principles to newspaper realities. As new technologies evolve, these principles can help guide editors to insure the credibility of the news and information they provide. Individual newspapers are encouraged to augment these APME guidelines more specifically to their own situations.

Responsibility

The good newspaper is fair, accurate, honest, responsible, independent and decent. Truth is its guiding principle.

It avoids practices that would conflict with the ability to report and present news in a fair, accurate and unbiased manner.

The newspaper should serve as a constructive critic of all segments of society. It should reasonably reflect, in staffing and coverage, its diverse constituencies. It should vigorously expose wrongdoing, duplicity or misuse of power, public or private. Editorially, it should advocate needed reform and innovation in the public interest. News sources should be disclosed unless there is a clear reason not to do so. When it is necessary to protect the confidentiality of a source, the reason should be explained.

The newspaper should uphold the right of free speech and freedom of the press and should respect the individual's right to privacy. The newspaper should fight vigorously for public access to news of government through open meetings and records.

Accuracy

The newspaper should guard against inaccuracies, carelessness, bias or distortion through emphasis, omission or technological manipulation.

It should acknowledge substantive errors and correct them promptly and prominently.

Integrity

The newspaper should strive for impartial treatment of issues and dispassionate handling of controversial subjects. It should provide a forum for the exchange of comment and criticism, especially when such comment is opposed to its editorial positions. Editorials and expressions of personal opinion by reporters and editors should be clearly labeled. Advertising should be differentiated from news.

The newspaper should report the news laws without regard for its own interests, mindful of the need to disclose potential conflicts. It should not give favored news treatment to advertisers or special-interest groups.

It should report matters regarding itself or its personnel with the same vigor and candor as it would other institutions or individuals. Concern for community, business or personal interests should not cause the newspaper to distort or misrepresent the facts.

The newspaper should deal honestly with readers and newsmakers. It should keep its promises.

The newspaper should not plagiarize words or images.

Independence

The newspaper and its staff should be free of obligations to news sources and newsmakers. Even the appearance of obligation or conflict of interest should be avoided.

Newspapers should accept nothing of value from news sources or others outside the profession. Gifts and free or reduced-rate travel, entertainment, products and lodging should not be accepted. Expenses in connection with news reporting should be paid by the newspaper. Special favors and special treatment for members of the press should be avoided.

Journalists are encouraged to be involved in their communities, to the extent that such activities do not create conflicts of interest. Involvement in politics, demonstrations and social causes that would cause a conflict of interest, or the appearance of such conflict, should be avoided.

Work by staff members for the people or institutions they cover also should be avoided.

Financial investments by staff members or other outside business interests that could create the impression of a conflict of interest should be avoided.

Stories should not be written or edited primarily for the purpose of winning awards and prizes. Self-serving journalism contests and awards that reflect unfavorably on the newspaper or the profession should be avoided.

IABC Code of Ethics for Professional Communicators

Preface

Because hundreds of thousands of business communicators worldwide engage in activities that affect the lives of millions of people, and because this power carries with it significant social responsibilities, the International Association of Business Communicators developed the Code of Ethics for Professional Communicators.

The Code is based on three different yet interrelated principles of professional communication that apply throughout the world.

These principles assume that just societies are governed by a profound respect for human rights and the rule of law; that ethics, the criteria for determining what is right and wrong, can be agreed upon by members of an organization; and, that understanding matters of taste requires sensitivity to cultural norms.

These principles are essential:

- Professional communication is legal.
- Professional communication is ethical.
- Professional communication is in good taste.

Recognizing these principles, members of IABC will:

- engage in communication that is not only legal but also ethical and sensitive to cultural values and beliefs;
- engage in truthful, accurate and fair communication that facilitates respect and mutual understanding; and,
- adhere to the following articles of the IABC Code of Ethics for Professional Communicators.

Because conditions in the world are constantly changing, members of IABC will work to improve their individual competence and to increase the body of knowledge in the field with research and education.

Articles

1. Professional communicators uphold the credibility and dignity of their profession by practicing honest, candid and timely communication and by fostering the free flow of essential information in accord with the public interest.

2. Professional communicators disseminate accurate information and promptly correct any erroneous communication for which they may be responsible.
3. Professional communicators understand and support the principles of free speech, freedom of assembly, and access to an open marketplace of ideas; and, act accordingly.
4. Professional communicators are sensitive to cultural values and beliefs and engage in fair and balanced communication activities that foster and encourage mutual understanding.
5. Professional communicators refrain from taking part in any undertaking which the communicator considers to be unethical.
6. Professional communicators obey laws and public policies governing their professional activities and are sensitive to the spirit of all laws and regulations and, should any law or public policy be violated, for whatever reason, act promptly to correct the situation.
7. Professional communicators give credit for unique expressions borrowed from others and identify the sources and purposes of all information disseminated to the public.
8. Professional communicators protect confidential information and, at the same time, comply with all legal requirements for the disclosure of information affecting the welfare of others.
9. Professional communicators do not use confidential information gained as a result of professional activities for personal benefit and do not represent conflicting or competing interests without written consent of those involved.
10. Professional communicators do not accept undisclosed gifts or payments for professional services from anyone other than a client or employer.
11. Professional communicators do not guarantee results that are beyond the power of the practitioner to deliver.
12. Professional communicators are honest not only with others but also, and most importantly, with themselves as individuals; for a professional communicator seeks the truth and speaks that truth first to the self.

Enforcement and Communication of the IABC Code for Professional Communicators

IABC fosters compliance with its Code by engaging in global communication campaigns rather than through negative sanctions. However, in keeping with the sixth article of the IABC Code, members of IABC who are found guilty by an appropriate governmental agency or judicial body of violating laws and public policies governing their professional activities may have their membership terminated by the IABC executive board following procedures set forth in the association's bylaws.

IABC encourages the widest possible communication about its Code.
The IABC Code of Ethics for Professional Communicators is published in several languages and is freely available to all: Permission is hereby granted to any individual or organization wishing to copy and incorporate all or part of the IABC Code into personal and corporate codes, with the understanding that appropriate credit be given to IABC in any publication of such codes.

The IABC Code is published in the association's annual directory, The WorldBook of IABC Communicators. The association's monthly magazine, Communication World, publishes periodic articles dealing with ethical issues. At least one session at the association's annual conference is devoted to ethics. The international headquarters of IABC, through its professional development activities, encourages and supports efforts by IABC student chapters, professional chapters, and districts/regions to conduct meetings and workshops devoted to the topic of ethics and the IABC Code. New and renewing members of IABC sign the following statement as part of their application: "I have reviewed and understand the IABC Code of Ethics for Professional Communicators."

As a service to communicators worldwide, inquiries about ethics and questions or comments about the IABC Code may be addressed to members of the IABC Ethics Committee. The IABC Ethics Committee is composed of at least three accredited members of IABC who serve staggered three-year terms. Other IABC members may serve on the committee with the approval of the IABC executive committee. The functions of the Ethics Committee are to assist with professional development activities dealing with ethics and to offer advice and assistance to individual communicators regarding specific ethical situations.

While discretion will be used in handling all inquiries about ethics, absolute confidentiality cannot be guaranteed. Those wishing more information about the IABC Code or specific advice about ethics are encouraged to contact IABC World Headquarters (One Hallidie Plaza, Suite 600, San Francisco, CA 94102 USA; phone, 415–544–4700; fax, 415–544–4747).

NPPA Code of Ethics

(Published October 26, 1999; Last Updated October 26, 1999)

Statement of Purpose

The National Press Photographers Association, a professional society dedicated to the advancement of photojournalism, acknowledges concern and respect for the public's natural-law, right to freedom in searching for the truth and the right to be informed truthfully and completely about public events and the world in which we live. NPPA believes that no report can be complete if it is possible to enhance and clarify the meaning of the words. We believe that pictures, whether used to depict news events as they actually happen, illustrate news that has happened, or to help explain anything of public interest, are indispensable means of keeping people accurately informed, that they help all people, young and old, to better understand any subject in the public domain. NPPA recognizes and acknowledges that photojournalists should at all times maintain the highest standards of ethical conduct in serving the public interest.

Code of Ethics

1. The practice of photojournalism, both as a science and art, is worthy of the very best thought and effort of those who enter into it as a profession.
2. Photojournalism affords an opportunity to serve the public that is equaled by few other vocations and all members of the profession should strive by example and influence to maintain high standards of ethical conduct free of mercenary considerations of any kind.
3. It is the individual responsibility of every photojournalist at all times to strive for pictures that report truthfully, honestly and objectively.
4. As journalists, we believe that credibility is our greatest asset. In documentary photojournalism, it is wrong to alter the content of a photograph in any way (electronically or in the darkroom) that deceives the public. We believe the guidelines for fair and accurate reporting should be the criteria for judging what may be done electronically to a photograph.
5. Business promotion in its many forms is essential but untrue statements of any nature are not worthy of a professional photojournalist and we severely condemn any such practice.
6. It is our duty to encourage and assist all members of our profession, individually and collectively, so that the quality of photojournalism may constantly be raised to higher standards.
7. It is the duty of every photojournalist to work to preserve all freedom-of-the-press rights recognized by law and to work to protect and expand freedom-of-access to all sources of news and visual information.
8. Our standards of business dealings, ambitions and relations shall have in them a note of sympathy for our common humanity and shall always require us to take into consideration our highest duties as members of society. In every situation in our business life, in every responsibility that comes before us, our chief thought shall be to fulfill that responsibility and discharge that duty so that when each of us is finished we shall have endeavored to lift the level of human ideals and achievement higher than we found it.

No Code of Ethics can prejudge every situation, thus common sense and good judgment are required in applying ethical principles.

RTNDA Code of Ethics and Professional Conduct

(Published December 09, 2002; Last Updated December 09, 2002)

The Radio-Television News Directors Association, wishing to foster the highest professional standards of electronic journalism, promote public understanding of and confidence in electronic journalism, and strengthen principles of journalistic freedom to gather and disseminate information, establishes this Code of Ethics and Professional Conduct.

Preamble

Professional electronic journalists should operate as trustees of the public, seek the truth, report it fairly and with integrity and independence, and stand accountable for their actions.

PUBLIC TRUST: Professional electronic journalists should recognize that their first obligation is to the public.

Professional electronic journalists should:

- Understand that any commitment other than service to the public undermines trust and credibility.
- Recognize that service in the public interest creates an obligation to reflect the diversity of the community and guard against oversimplification of issues or events.
- Provide a full range of information to enable the public to make enlightened decisions.
- Fight to ensure that the public's business is conducted in public.

TRUTH: Professional electronic journalists should pursue truth aggressively and present the news accurately, in context, and as completely as possible.

Professional electronic journalists should:

- Continuously seek the truth.
- Resist distortions that obscure the importance of events.
- Clearly disclose the origin of information and label all material provided by outsiders.

Professional electronic journalists should not:

- Report anything known to be false.
- Manipulate images or sounds in any way that is misleading.
- Plagiarize.
- Present images or sounds that are reenacted without informing the public.

FAIRNESS: Professional electronic journalists should present the news fairly and impartially, placing primary value on significance and relevance. Professional electronic journalists should:

- Treat all subjects of news coverage with respect and dignity, showing particular compassion to victims of crime or tragedy.
- Exercise special care when children are involved in a story and give children greater privacy protection than adults.
- Seek to understand the diversity of their community and inform the public without bias or stereotype.
- Present a diversity of expressions, opinions, and ideas in context.
- Present analytical reporting based on professional perspective, not personal bias.
- Respect the right to a fair trial.

INTEGRITY: Professional electronic journalists should present the news with integrity and decency, avoiding real or perceived conflicts of interest, and respect the dignity and intelligence of the audience as well as the subjects of news. Professional electronic journalists should:

- Identify sources whenever possible. Confidential sources should be used only when it is clearly in the public interest to gather or convey important information or when a person providing information might be harmed. Journalists should keep all commitments to protect a confidential source.
- Clearly label opinion and commentary.
- Guard against extended coverage of events or individuals that fails to significantly advance a story, place the event in context, or add to the public knowledge.
- Refrain from contacting participants in violent situations while the situation is in progress.
- Use technological tools with skill and thoughtfulness, avoiding techniques that skew facts, distort reality, or sensationalize events.
- Use surreptitious newsgathering techniques, including hidden cameras or microphones, only if there is no other way to obtain stories of significant public importance and only if the technique is explained to the audience.
- Disseminate the private transmissions of other news organizations only with permission.

Professional electronic journalists should not:

- Pay news sources who have a vested interest in a story.
- Accept gifts, favors, or compensation from those who might seek to influence coverage.

- Engage in activities that may compromise their integrity or independence.

INDEPENDENCE: Professional electronic journalists should defend the independence of all journalists from those seeking influence or control over news content.

Professional electronic journalists should:

- Gather and report news without fear or favor, and vigorously resist undue influence from any outside forces, including advertisers, sources, story subjects, powerful individuals, and special interest groups.
- Resist those who would seek to buy or politically influence news content or who would seek to intimidate those who gather and disseminate the news.
- Determine news content solely through editorial judgment and not as the result of outside influence.
- Resist any self-interest or peer pressure that might erode journalistic duty and service to the public.
- Recognize that sponsorship of the news will not be used in any way to determine, restrict, or manipulate content.
- Refuse to allow the interests of ownership or management to influence news judgment and content inappropriately.
- Defend the rights of the free press for all journalists, recognizing that any professional or government licensing of journalists is a violation of that freedom.

ACCOUNTABILITY: Professional electronic journalists should recognize that they are accountable for their actions to the public, the profession, and themselves.

Professional electronic journalists should:

- Actively encourage adherence to these standards by all journalists and their employers.
- Respond to public concerns. Investigate complaints and correct errors promptly and with as much prominence as the original report.
- Explain journalistic processes to the public, especially when practices spark questions or controversy.
- Recognize that professional electronic journalists are duty-bound to conduct themselves ethically.
- Refrain from ordering or encouraging courses of action that would force employees to commit an unethical act.
- Carefully listen to employees who raise ethical objections and create environments in which such objections and discussions are encouraged.

- Seek support for and provide opportunities to train employees in ethical decision-making.

In meeting its responsibility to the profession of electronic journalism, RTNDA has created this code to identify important issues, to serve as a guide for its members, to facilitate self-scrutiny, and to shape future debate.

SABEW Code of Ethics

Statement of Purpose: It is not enough that we be incorruptible and act with honest motives. We must conduct all aspects of our lives in a manner that averts even the appearance of conflict of interest or misuse of the power of the press.

A business, financial and economics writer should:

1. Recognize the trust, confidence and responsibility placed in him or her by the publication's readers and do nothing to abuse this obligation. To this end, a clear-cut delineation between advertising and editorial matters should be maintained at all times.
2. Avoid any practice which might compromise or appear to compromise his objectivity or fairness. He or she should not let any personal investments influence what he or she writes. On some occasions, it may be desirable for him or her to disclose his or her investment positions to a superior.
3. Avoid active trading and other short-term profit-seeking opportunities. Active participation in the markets which such activities require is not compatible with the role of the business and financial journalist as disinterested trustee of the public interest.
4. Not take advantage in his or her personal investing of any inside information and be sure any relevant information he or she may have is widely disseminated before he buys or sells.
5. Make every effort to insure the confidentiality of information held for publication to keep such information from finding its way to those who might use it for gain before it becomes available to the public.
6. Accept no gift, special treatment or any other thing of more than token value given in the course of his professional activities. In addition, he or she will accept no out-of-town travel paid for by anyone other than his or her employer for the ostensible purpose of covering or backgrounding news. Free-lance writing opportunities and honoraria for speeches should be examined carefully to assure that they are not in fact disguised gratuities. Food and refreshments of ordinary value may be accepted where necessary during the normal course of business.
7. Encourage the observance of these minimum standards by all business writers.

Addendum to Code of Ethics

Guidelines To Insure Editorial Integrity Of Business News Coverage:

1. A clear-cut delineation between advertising and editorial matters should be maintained at all times.
2. Material produced by an editorial staff or news service should be used only in sections controlled by editorial departments.
3. Sections controlled by advertising departments should be distinctly different from news sections in typeface, layout and design.
4. Promising a story in exchange for advertising is unethical.
5. Publishers, broadcasters and top newsroom editors should establish policies and guidelines to protect the integrity of business news coverage.

Cautions on Use of Non-Journalists With Conflicts of Interest in the Subject Matter:

Using articles or columns written by non-journalists is potentially deceptive and poses inherent conflicts of interest that editors should guard against. This does not apply to clearly labeled op-ed or viewpoint sections or "Letters to the Editor."

Gannett Newspaper Division

I. Principles of Ethical Conduct for Newsrooms

WE ARE COMMITTED TO:

Seeking and reporting the truth in a truthful way

- We will dedicate ourselves to reporting the news accurately, thoroughly and in context.
- We will be honest in the way we gather, report and present news.
- We will be persistent in the pursuit of the whole story.
- We will keep our word.
- We will hold factual information in opinion columns and editorials to the same standards of accuracy as news stories.
- We will seek to gain sufficient understanding of the communities, individuals and stories we cover to provide an informed account of activities.

Serving the public interest

- We will uphold First Amendment principles to serve the democratic process.
- We will be vigilant watchdogs of government and institutions that affect the public.

- We will provide the news and information that people need to function as effective citizens.
- We will seek solutions as well as expose problems and wrongdoing.
- We will provide a public forum for diverse people and views.
- We will reflect and encourage understanding of the diverse segments of our community.
- We will provide editorial and community leadership.
- We will seek to promote understanding of complex issues.

Exercising fair play

- We will treat people with dignity, respect and compassion.
- We will correct errors promptly.
- We will strive to include all sides relevant to a story and not take sides in news coverage.
- We will explain to readers our journalistic processes.
- We will give particular attention to fairness in relations with people unaccustomed to dealing with the press.
- We will use unnamed sources as the sole basis for published information only as a last resort and under specific procedures that best serve the public's right to know.
- We will be accessible to readers.

Maintaining independence

- We will remain free of outside interests, investments or business relationships that may compromise the credibility of our news report.
- We will maintain an impartial, arm's length relationship with anyone seeking to influence the news.
- We will avoid potential conflicts of interest and eliminate inappropriate influence on content.
- We will be free of improper obligations to news sources, newsmakers and advertisers.
- We will differentiate advertising from news.

Acting with integrity

- We will act honorably and ethically in dealing with news sources, the public and our colleagues.
- We will obey the law.
- We will observe common standards of decency.
- We will take responsibility for our decisions and consider the possible consequences of our actions.
- We will be conscientious in observing these Principles.
- We will always try to do the right thing.

II. Protecting the Principles

No statement of principles and procedures can envision every circumstance that may be faced in the course of covering the news. As in the United States Constitution, fundamental principles sometimes conflict. Thus, these recommended practices cannot establish standards of performance for journalists in every situation.

Careful judgment and common sense should be applied to make the decisions that best serve the public interest and result in the greatest good. In such instances, journalists should not act unilaterally. The best decisions are obtained after open-minded consultations with appropriate colleagues and superiors—augmented, when necessary, by the advice of dispassionate outside parties, such as experts, lawyers, ethicists, or others whose views in confidence may provide clarity in sorting out issues.

Here are some recommended practices to follow to protect the Principles. This list is not all-inclusive. There may be additional practices—implicit in the Principles or determined within individual newsrooms—that will further ensure credible and responsible journalism.

Ensuring the Truth Principle

"Seeking and reporting the truth in a truthful way" includes, specifically:

- We will not lie.
- We will not misstate our identities or intentions.
- We will not fabricate.
- We will not plagiarize.
- We will not alter photographs to mislead readers.
- We will not intentionally slant the news.

Using unnamed sources

The use of unnamed sources in published stories should be rare and only for important news. Whenever possible, reporters should seek to confirm news on the record. If the use of unnamed sources is required:

- Use as sources only people who are in a position to know.
- Corroborate information from an unnamed source through another source or sources and/or by documentary information. Rare exceptions must be approved by the editor.
- Inform sources that reporters will disclose sources to at least one editor. Editors will be bound by the same promise of confidentiality to sources as are reporters.
- Hold editors as well as reporters accountable when unnamed sources are used. When a significant story to be published relies on a source who will not be named, it is the responsibility of the senior

news executive to confirm the identity of the source and to review the information provided. This may require the editor to meet the source.

- Make clear to the reporters and to sources that agreements of confidentiality are between the newspaper and the sources, not just between the reporter and the sources. The newspaper will honor its agreements with sources. Reporters should make every effort to clear such confidentiality agreements with the editors first. Promises of confidentiality made by reporters to sources will not be overridden by the editors; however, editors may choose not to use the material obtained in this fashion.
- Do not allow unnamed sources to take cheap shots in stories. It is unfair and unprofessional.
- Expect reporters and editors to seek to understand the motivations of a source and take those into account in evaluating the fairness and truthfulness of the information provided.
- Make clear to sources the level of confidentiality agreed to. This does not mean each option must be discussed with the source, but each party should understand the agreement. Among the options are:
  a. The newspaper will not name them in the article;
  b. The newspaper will not name them unless a court compels the newspaper to do so;
  c. The newspaper will not name them under any circumstances.

All sources should be informed that the newspaper will not honor confidentiality if the sources have lied or misled the newspaper.

- Make sure both sides understand what is being agreed to. For example:
  a. Statements may be quoted directly or indirectly and will be attributed to the source. This is sometimes referred to as "on the record."
  b. The information may be used in the story but not attributed to the source. This is sometimes referred to as "not for attribution" or "for background."
  c. The information will not be used in the story unless obtained elsewhere and attributed to someone else. This is sometimes referred to as "off the record."
- Describe an unnamed source's identity as fully as possible (without revealing that identity) to help readers evaluate the credibility of what the source has said or provided.
- Do not make promises you do not intend to fulfill or may not be able to fulfill.
- Do not threaten sources.

Handling the wires

These Principles are intended to provide front-line guidance for locally generated material. Wire-service material already has been edited professionally. Gannett News Service observes these same Principles. The Associated Press has its own standards for the use of unnamed sources. Other wire-service standards may be lower. Additional scrutiny often is required, and further editing is encouraged. Ultimately, an editor must make a sound judgment about how to reconcile conflicts between wire-service and local-newsroom practices. Whenever possible, these Principles should prevail.

Being fair

Because of timeliness or unavailability, it is not always possible to include a response from the subject of an accusation in a news story. Nevertheless:

- We should make a good-faith effort to seek appropriate comment from the person (or organization) before publication.
- When that is not feasible, we should be receptive to requests for a response or try to seek a response for a follow-up story.
- Letters to the editor also may provide an appropriate means for reply.

Some public records will identify persons accused of wrongdoing. Publication of denials is not necessary in such circumstances.

Being independent

"Maintaining independence" helps establish the impartiality of news coverage. To clarify two points:

- News staff members are encouraged to be involved in worthwhile community activities, so long as this does not compromise the credibility of news coverage.
- When unavoidable personal or business interests could compromise the newspaper's credibility, such potential conflicts must be disclosed to one's superior and, if relevant, to readers.

Conducting investigative reporting

Aggressive, hard-hitting reporting is honorable and often courageous in fulfilling the press' First Amendment responsibilities, and it is encouraged. Investigative reporting by its nature raises issues not ordinarily faced in routine reporting. Here are some suggested procedures to follow when undertaking investigative reporting:

- Involve more than one editor at the early stages and in the editing of the stories.
- Question continually the premise of the stories and revise accordingly.

- Follow the practices outlined in the use of sources.
- Document the information in stories to the satisfaction of the senior editor.
- Have a "fresh read" by an editor who has not seen the material as you near publication. Encourage the editor to read it skeptically, then listen carefully to and heed questions raised about clarity, accuracy and relevance.
- Make certain that care, accuracy and fairness are exercised in headlines, photographs, presentation and overall tone.
- Evaluate legal and ethical issues fully, involving appropriate colleagues, superiors, lawyers or dispassionate outside parties in the editorial process. (For example, it may be helpful to have a technical story reviewed by a scientist for accuracy, or have financial descriptions assessed by an accountant, or consult an ethicist or respected outside editor on an ethical issue.)
- Be careful about trading information with sources or authorities, particularly if it could lead to an impression that you are working in concert against an individual or entity.

Editing skeptically

Editors are the gatekeepers who determine what will be published and what will not be. Their responsibility is to question and scrutinize, even when it is uncomfortable to do so. Here are some suggested practices that editors can follow:

- Take special care to understand the facts and context of the story.
- Guard against assumptions and preconceived notions—including their own.
- Ensure time and resources for sound editing. Nothing should be printed that has not been reviewed by someone else. When feasible, at least two editors should see stories before publication. Complex or controversial stories may require even more careful scrutiny.
- Consider involving an in-house skeptic on major stories—a contrarian who can play the role of devil's advocate.
- Challenge conventional wisdom.
- Heed their "gut instinct." Don't publish a story if it doesn't feel right. Check it further.
- Consider what may be missing from the story.
- Consider how others—especially antagonists or skeptical readers—may view the story. What questions would they ask? What parts would they think are unfair? Will they believe it?
- Be especially careful of stories that portray individuals purely as villains or heroes.

- Beware of stories that reach conclusions based on speculation or a pattern of facts.
- Protect against being manipulated by advocates and special interests.
- Consider these questions: "How do you know? How can you be sure? Where is the evidence? Who is the source? How does he or she know? What is the supporting documentation?"
- Watch carefully for red flags that give reason to be skeptical of news-gathering or editing conduct.
- Don't be stampeded by deadlines, unrealistic competitive concerns or peer pressure.

Ensuring accuracy

Dedication to the truth means accuracy itself is an ethical issue. Each news person has the responsibility to strive for accuracy at each step of the process.

- Be aware that information attributed to a source may not be factually correct.
- Be sure the person quoted is in a position to know.
- Be especially careful with technical terms, statistics, mathematical computations, crowd estimates and poll results.
- Consider going over all or portions of an especially complicated story with primary sources or with outside experts. However, do not surrender editorial control.
- Don't make assumptions. Don't guess at facts or spellings. Asking the person next to you is not "verification"—he or she could be wrong too.
- Improve note taking. Consider backing up your notes with a tape recorder when ethically and legally appropriate.
- Be wary of newspaper library clippings, which may contain uncorrected errors.
- Develop checklists of troublesome or frequently used names, streets, titles, etc.
- Understand the community and subject matter. Develop expertise in areas of specialized reporting.
- Reread stories carefully after writing, watching especially for errors of context and balance as well as for spelling and other basic mistakes.
- Use care in writing headlines. Do not stretch beyond the facts of the story.
- Follow a simple rule on the copy desk to double-check the accuracy of headlines: "Find the headline in the story." (For example, if the

headline says, "Three die in crash," go to the story and count the dead and be certain they died in the crash.)

- Consider using "accuracy checks" as an affirmative way to search out errors and monitor accuracy. (Accuracy checks are a process by which published stories are sent to sources or experts asking for comment on accuracy, fairness or other aspects.)

Correcting errors

When errors occur, the newspaper has an ethical obligation to correct the record and minimize harm.

- Errors should be corrected promptly. But first, a determination must be made that the fact indeed was in error and that the correction itself is fully accurate.
- Errors should be corrected with sufficient prominence that readers who saw the original error are likely to see the correction. This is a matter of the editor's judgment.
- Although it is wise to avoid repeating the error in the correction, the correction should have sufficient context that readers will understand exactly what is being corrected. ·
- Errors of nuance, context or tone may require clarifications, editor's notes, editor's columns or letters to the editor.
- When the newspaper disagrees with a news subject about whether a story contained an error, editors should consider offering the aggrieved party an opportunity to express his or her view in a letter to the editor.
- Corrections should be reviewed before publication by a senior editor who was not directly involved in the error. The editor should determine if special handling or outside counsel are required.
- Errors should be corrected whether or not they are called to the attention of the newspaper by someone outside the newsroom.
- Factual errors should be corrected in most cases even if the subject of the error does not want it to be corrected. The rationale for this is rooted in the Truth Principle. It is the newspaper's duty to provide accurate information to readers. An exception may be made —at the behest of the subject—when the correction of a relatively minor mistake would result in public ridicule or greater harm than the original error.
- Newsroom staffers should be receptive to complaints about inaccuracies and follow up on them.
- Newsroom staffers have a responsibility to alert the appropriate editor if they become aware of a possible error in the newspaper.

III. Reinforcing the Principles

Communicating standards

Editors have a responsibility to communicate these Principles to newsroom staff members and to the public. They should:

- Ensure that sound hiring practices are followed to build a staff of ethical and responsible journalists. Such practices include making reference checks and conducting sufficient interviewing and testing to draw reasonable conclusions about the individual's personal standards.
- Provide prospective hires with a copy of these Principles and make acceptance of them a condition of employment.
- Conduct staff training at least annually in the Principles of Ethical Conduct.
- Require staff members at the time of hire and each year thereafter to sign a statement acknowledging that they have read the Principles of Ethical Conduct and will raise any questions about them with their editors.
- Communicate these Principles to the public periodically.

Being accountable

Because these Principles embody the highest standards of professional conduct, the Gannett Newspaper Division is committed to their adherence. They have been put in writing specifically so that members of every Gannett Newspaper Division newsroom know what the Division stands for and what is expected of them. The public will know, too.

Notes

Introduction

1 *American Heritage Dictionary,* s.v. "journalism."
2 "Definition of 'Journalism' Is a Sticky Question: And an Even Stickier Question—Do We Need a Definition?" *Quill,* May 2002, p. 1.

Chapter 1

1 Ron F. Smith, *Groping for Ethics in Journalism,* 4th ed. (Ames: Iowa State Press, 1999), p. 8.
2 James Madison, Letter to W.T. Barry, August 4, 1822, in Gaillard Hunt (Ed.), 9 *James Madison's Writings* 103 (1910).
3 "Codes of Ethics and Beyond," PoynterOnline, April 1, 1999. Retrieved from the World Wide Web: http://www.poynter.org/content/content_view.asp?id=5522).
4 E-mail interview with Gary Hill, December 16, 2002.
5 www.cnet.com/aboutcnet/0,10000,0-13611-7-920000,00.html.
6 Rebecca Blood, *The Weblog Handbook: Practical Advice on Creating and Maintaining Your Blog* (Cambridge, MA: Perseus, 2002).
7 A fuller discussion of each of those principles is available on the Web as an excerpt from the book at http://rebeccablood.net/handbook/excerpts/weblog_ethics.html, pp. 117–121.
8 Robert Kane, *Through the Moral Maze: Searching for Absolute Values in a Pluralistic World* (New York: Longman, 1994). In Richard L. Johannesen, *Ethics in Human Communication,* 5th ed. (Prospect Heights, IL: Waveland, 2002), p. 230.

9 Johannesen, p. 233.
10 An international media code of ethics of sorts was developed in Paris in 1983, called the "International Principles of Professional Ethics in Journalism." However, this statement emerged from a very contentious attempt by Third World countries under the auspices of UNESCO to create what was called a "New World Information and Communications Order (NWICO)," which was designed to place limits on media imports. NWICO was not at all well received in the United States or the United Kingdom and was seen by some as an attempt to use the government to control press freedoms.
11 All codes are quoted from the International Journalists' Network Web site at www.ijnet.org/code.html.
12 International Journalists' Network, http://www.ijnet.org/5129.html.
13 International Journalists' Network, www.ijnet.org/Code_of_Ethics2/Finland__Union_of_Journalists_in_Finland.html.

Chapter 2

1 Kenneth Starck, "What's Right/Wrong with Journalism Ethics Research?" *Journalism Studies*, Vol. 2, No. 1 (2001), pp. 134–135.
2 Neil Postman, *Amusing Ourselves to Death* (New York: Penguin, 1985), p. 34.
3 Mitchell Stephens, *A History of News: From the Drum to the Satellite* (New York: Viking, 1988), pp. 185–193.
4 Mitchell Stephens, "A Call for an International History of Journalism." Article posted on Stephens' Web site at NYU. See www.nyu.edu/classes/stephens/International%20History%20page.htm.
5 See James W. Carey, *Communication as Culture* (New York: Routledge, 1989), pp. 142–172, and Harold Innis, *The Bias of Communication* (Toronto: University of Toronto Press, 1997, originally published 1951).
6 Carey, p. 82.
7 Jurgen Habermas, *The Structural Transformation of the Public Sphere: An Inquiry into a Category of Bourgeois Society* (Cambridge: MIT Press, 1989).
8 Patricia Aufderheide, "Telecommunications and the Public Interest." In E. Barnouw (Ed.), *Conglomerates and the Media* (New York: New Press, 1997), p. 167.
9 Habermas.
10 Mark Poster, "The Net as Public Sphere." In David Crowley and Paul Heyer (Eds.), *Communication in History* (New York: Longman, 1999), p. 335.
11 Stephens, p. 201.
12 Innis, p. 156.
13 Darrell West, *The Rise and Fall of the Media Establishment* (Boston: Bedford/St. Martins, 2001), chap. 2.

14 West, pp. 10–12.

15 Aufderheide, p. 167.

16 West, p. 19.

17 Alexis de Tocqueville, *Democracy in America* (New York: Penguin, 1965), p. 93.

18 de Toqueville, p. 94.

19 Stephens, p. 202.

20 Quoted in West, p. 19.

21 Innis, p. 160.

22 Innis, p. 161.

23 Stephens, p. 205.

24 Stephens, p. 205.

25 Stephens, pp. 242–247.

26 Stephens, p. 208.

27 Crowley & Heyer, p. 132.

28 Postman pp. 64–69; James W. Carey, "Time, Space, and the Telegraph." In David Crowley and Paul Heyer (Eds.), *Communication in History* (New York: Longman, 1999), pp. 135–140.

29 Carey, "Time, Space, and the Telegraph," p. 139.

30 West, p. 27.

31 Postman, p. 69.

32 Carey, *Communication as Culture,* p. 168, emphasis added.

33 Michael Schudson, *Discovering the News: A Social History of American Newspaper* (New York: Basic Books, 1978), p. 143; West pp.27–45.

34 Ted Curtis Smythe, "The Diffusion of the Urban Daily, 1850–1900," *Journalism History*, Vol. 28, No. 2 (Summer 2002) pp. 73–85.

35 William Leiss, Stephen Kline, and Sut Jhally, "Advertising, Consumers, and Culture." In David Crowley and Paul Heyer (Eds.), *Communication in History* (New York: Longman, 1999), p. 210.

36 Edward S. Herman and Noam Chomsky, *Manufacturing Consent: The Political Economy of Mass Media* (New York: Pantheon, 1988), p. 14.

37 Michael Schudson, "The New Journalism." In David Crowley and Paul Heyer (Eds.), *Communication in History* (New York: Longman, 1999), pp. 141–142.

38 Stephens, pp. 208–211.

39 Stephens, p. 210.

40 Schudson, "The New Journalism," pp. 147–148.

41 Schudson, "The New Journalism," p. 147.

42 Schudson, "The New Journalism," p. 147.

43 Noam Chomsky, paraphrased from interview with Chomsky in "Manufacturing Consent: Noam Chomsky and the Media," documentary film, produced by Marc Achbar and Peter Wintonnick, Zeitgeist home video release, 1994.

44 Leiss, Kline, & Jhally, p. 210; John Streck, "Pulling the Plug on Electronic Town Meetings: Participatory Democracy and the Reality of Usenet." In Chris Toulouse and Timothy W. Luke (Eds.), *The Politics of Cyberspace* (New York: Routledge, 1998), p. 24.

45 Streck, p. 24.

46 Habermas, p. 155.

47 Innis, p. 182.

48 George Seldes, *Lords of the Press* (New York: Julian Messner, Inc., 1938), p. 78.

49 Innis, p. 186.

50 West, p. 45.

51 C. Wright Mills, *The Power Elite* (New York: Oxford University Press, 1956), pp. 303–304.

52 Starck, p. 134.

53 Starck, p. 135.

54 Stephens, pp. 248–250.

55 Stephens, p. 250.

56 See Starck, p. 135, and Robert W. McChesney and Ben Scott, "Upton Sinclair and the Contradictions of Capitalist Journalism," *Monthly Review*, Vol. 54, No. 1 (May 2002). Accessed online November 23, 2002, at www.monthlyreview.org/0502rwmscott.htm.

57 Quoted in West, p. 50.

58 West, p. 50.

59 Stephens, p. 229.

60 West, p. 54.

61 Marcus Errico, "The Evolution of the Summary News Lead," *Media History Monographs*, Vol. 1, No. 1 (1997/1998).

62 Walter Lippmann, *Public Opinion* (New York: Free Press, 1965 edition).

63 See Carey, *Communication as Culture*, pp. 74–82, and James Fallows, *Breaking the News: How the Media Undermine American Democracy* (New York: Vintage, 1997), chap. 6.

64 John Dewey, *The Public and Its Problems* (Athens: University of Ohio Press, 1983, originally published in 1927), p. 208.

65 Carey, *Communication as Culture*, p. 82.

66 Stephen Bates, *Realigning Journalism with Democracy: The Hutchins Commission, Its Times, and Ours* (Washington, DC: Annenberg Washington Program in Communications Policy Studies of Northwestern University, 1995), www.annenberg.nwu.edu/pubs/hutchins/.

67 J. Douglas Tarpley, "Journalism: Truth and the 21st Century," Gegrapha .org, August 11, 1998, http://www.gegrapha.org/resources/tarpley1.htm.

68 William Stott, "Documenting Media." In David Crowley and Paul Heyer (eds.), *Communication in History* (New York: Longman, 1999); Stuart Ewen, *PR! A Social History of Spin* (New York: Basic Books, 1996), pp. 236–237.

69 Robert McChesney, *Rich Media, Poor Democracy: Communications Politics in Dubious Times* (Champaign: University of Illinois Press, 1999), chap. 4.

70 West, p. 59.

71 Seldes.

72 See Bates for a detailed summary of the commission's deliberations.

73 See Bates.

74 Robert D. Leigh, ed., *A Free and Responsible Press* (Chicago: University of Chicago Press, 1947), p. 11.

75 Paraphrased from the Hutchins report, pp. 17–28: Commission on Freedom of the Press, *A Free and Responsible Press: A General Report on Mass Communications—Newspapers, Radio, Motion Pictures, Magazines, and Books* (Chicago: University of Chicago Press, 1947).

76 Paraphrased from Bates.

77 Quoted in Bates.

78 The *Chicago Tribune* quoted in Bates.

79 Paul J. Thompson, chair of Journalism at the University of Texas, quoted in Bates.

80 See Bates' postcript.

81 See McChesney, pp.63–77, and Ben Bagdikian, *The Media Monopoly,* 5th ed. pp. xxx–xxxi(Boston: Beacon Press, 1997).

82 Ewen, p. 386.

83 Quoted in Herbert I. Schiller, *Culture, Inc.* (New York: Oxford University Press, 1989), p. 122.

84 West, p. 58.

85 West, appendix A.2, p. 131.

86 Jeffrey Scheuer, *The Soundbite Society* (New York: Routledge, 2001), p. 76.

87 Philip Meyer, "Public Journalism and the Problem of Objectivity." This article is based on a talk given to the Investigative Reporters and Editors conference on computer-assisted reporting in Cleveland, OH, September 1995. Online www.unc.edu/~pmeyer/ire95pj.htm.

88 Pierre Bourdieu, *On Television* (New York: New Press, 1998).

89 Postman, chap. 6.

90 Jeffery Scheuer, *The Soundbite Society* (New York: Routledge, 2001), p. 81

91 The Lear Center Local News Archive, "Political Ads Dominate Local TV News Coverage," November 1, 2002, www.localnewsarchive.org/.

92 Scheuer, p. 84.

93 Michael Schudson, "Was There Ever a Public Sphere?" In Craig Calhoun (Ed.), *Habermas and the Public Sphere* (Cambridge: MIT Press, 1999), p. 150.

94 C. Edwin Baker, "Market Threats to Press Freedom." Address to the 15th Annual Symposium on American Values, Angelo State University, San Angelo, TX, October 12, 1998. www.angelo.edu/events/university_symposium/1998/baker.htm.

95 Eric Hobsbawm, *Hobsbawm on History* (New York: New Press, 1997), p. xiii.

96 Meyer.

97 Robert Putnam, "Bowling Alone: America's Declining Social Capital," *Journal of Democracy,* Vol. 6, No. 1 (January 1995), pp. 65–78.

98 Pew Research Center for the People and the Press, "Self Censorship: How Often and Why?" April 30, 2000. Online at http://people-press.org/reports/display.php3?ReportID=39.

99 Tom Rosenstiel and Dave Iverson, "Politics and TV Can Mix," *Los Angeles Times*, October 15, 2002, and PEJ Web site at www.journalism.org/resources/publications/articles/tvpolitics.asp.

100 Jay Rosen, *What Are Journalists For?* (New Haven: Yale University Press, 1999).

101 Fallows, pp. 255–257, and Ron F. Smith, *Groping for Ethics in Journalism* (Ames: Iowa State Press, 1999), pp. 313–314.

102 Jay Rosen, "What Is Public Journalism? A Brief Description." Article posted on Jay Rosen's NYU Web site August 1999, www.nyu.edu/gsas/dept/journal/Faculty/bios/rosen/public_journalism.htm. Accessed November 27, 2002.

103 Smith, p. 314.

104 Smith, p. 314.

105 McChesney, pp. 300–301.

106 Rosen, "What Is Public Journalism?"

107 Fallows, p. 260.

Chapter 3

1 *Branzburg v. Hayes*, 408 U.S. 665.

2 Gina Barton, "What Is a Journalist?" *Quill,* May 2002.

3 *Von Bulow v. Von Bulow*, 811 F.2d 136.

4 *In re Madden*, 151 F.3d 125.

5 *Lovell v. Griffin*, 303 U.S. 444.

6 *In re Madden*.

7 Barton.

8 Barton.

9 Barton.

10 To keep up to date on the latest legal cases involving journalism and the Internet, the authors recommend the legal section of the Online Journalism Review (www.ojr.org/), as well as the Reporters Committee for the Freedom of the Press at www.rcfp.org/ and the Society for Professional Journalists at www.spj.org.

11 "What Is Journalism, Who Is a Journalist?" Committee of Concerned Journalists Forum, Chicago, IL, November 6, 1997. Transcript online at www.journalism.org/resources/education/forums/ccj/forum1/who.asp.

12 "The Constitution Had It Right," *The Register-Guard,* September 21, 2000.

13 Poynter Institute's Online News discussion list, posted August 3, 2002.

14 Bill Kovach and Tom Rosenstiel, *Warp Speed: America in the Age of Mixed Media Culture* (Washington, DC: Century Foundation Press, 1999), p. 7.

15 Kovach & Rosenstiel, p. 7.

16 Kovach & Rosenstiel, p. 2.

17 An extended discussion of these principles can be found in Bill Kovach and Tom Rosenstiel, *The Elements of Journalism: What Newspeople Should Know and the Public Should Expect* (New York: Crown, 2001).

18 Barton.

19 Robert McChesney, *Rich Media, Poor Democracy: Communications Politics in Dubious Times* (Champaign: University of Illinois Press, 1999), pp. 48–62.

20 Robert McChesney and Ben Scott, "Upton Sinclair and the Contradiction of Capitalist Journalism," *Monthly Review*, Vol. 54, No. 1 (May 2002).

21 Jon Katz, "What's Really Happening with 'Struggling' Online News," Freedom Forum online, June 21, 2000.

22 Source for all of Miller's comments: personal interview conducted on March 19, 2002.

23 "Interview with Rusty Foster," Dotcom Scoop, January 28, 2002.

24 "Open-Source Journalism," Salon.com, October 9, 1999.

25 See George Albert Gladney and James D. Ivory, "Attitudes of Relational Engagement in Cyberspace: Uncovering Monologic Potential and Growth." Paper presented at UDC conference, State College, PA, October 11, 2002.

26 "Interview with Rusty Foster."

27 Katz.

28 "Interview with Rusty Foster."

29 "Much Ado About Blogging," Salon.com, May 10, 2002, original emphasis.

30 "Much Ado About Blogging."

31 "Much Ado About Blogging."

32 "J-blogs: Weblogs, Journalism, Sources, Witnessing," from Nublog, March, 5, 2002.

33 "Columbia's J-School Needs to Consider Trollopian Retooling," *New York Observer* online, August 26, 2002.

34 "News for the People, by the People," Online Journalism Review, May 16, 2002.

35 "Blogging Goes Legit, Sort Of," Wired.com, June 6, 2002.

36 "Old and New Journalism Together." From Dan Gillmor's "e-journal," May 1, 2002.

37 "Much Ado About Blogging."

38 "When Bloggers Commit Journalism," Online Journalism Review, September 24, 2002. Retrieved from the World Wide Web: www.ojr.org/ojr/lasica/1032910520.php.

39 See Mitch Ratcliffe's blog at www.ratcliffe.com/bizblog/2002/10/12.html#a33.

40 Richard L. Johannesen, *Ethics in Human Communication*, 5th ed. (Prospect Heights, IL: Waveland, 2002), p. 7.

41 See, for example, Dan Gillmor, "Accessing a Whole New World via Multimedia Phones," *Mercury News,* December 8, 2002. Retrieved from the World Wide Web: www.siliconvalley.com/mld/siliconvalley/.

42 Online Journalism Review, July 11, 2002, emphasis added.

43 Mark Deuze, "Online Journalism: Modeling the First Generation of News Media on the World Wide Web," *First Monday,* Vol. 6, No. 10 (October 2001).

44 "Interview with Rusty Foster."

Chapter 4

1 Jeffrey Rosen, *The Unwanted Gaze* (New York: Random House, 2000), p. 19.

2 "Privacy and Human Rights Summary 2002," EPIC/Privacy International, p. 5.

3 "Privacy and Human Rights," p. 5.

4 See Article 12 in the UD at www.un.org/Overview/rights.html.

5 "Privacy and Human Rights," p. 4.

6 "Why Privacy?" *Boston Globe*, September 2, 2001. Accessed from www.opengovva.org/why-privacy.html.

7 "The Right to Privacy," *Harvard Law Review,* Vol. IV, No. 5 (December 15, 1890).

8 See *Database Nation: The Death of Privacy in the 21st Century* (Sebastopol, CA: O'Reilly and Associates, 2000).

9 "The Right to Privacy."

10 "The Right to Privacy."

11 *Sidis v. F-R Pub. Corp.,* 1940.

12 *Olmstead v. U.S.,* 277 U.S. 438 1928.

13 *Katz v. U.S.,* 386 U.S. 954 1967.

14 See, for example, Garfinkel.

15 Adapted from list on EPIC Web site. See www.epic.org/privacy/consumer/code_fair_info.html.

16 "Privacy and Human Rights," p. 383.

17 "FBIs Reliance on the Private Sector," *Wall Street Journal,* April 13, 2001.

18 "FBIs Reliance."

19 Tim Berners-Lee, *Weaving the Web: The Original Design and Ultimate Destiny of the World Wide Web by Its Inventor* (San Francisco: Harper Collins, 1999), p. 18.

20 Berners-Lee, p. 18.

21 "Dirty Laundry, Online for All to See," NYTimes.com, September 2, 2002.

22 "Balancing Privacy and the Right to Know," *San Jose Mercury News,* December 8, 2001, p. 1A.

23 "Balancing Privacy."

24 "Dirty Laundry."

25 "Dirty Laundry."

26 "Privacy and Human Rights," p. 390.

27 "Librarians Under Siege," *The Nation,* August 5, 2002.

28 Paul Sholtz, "Economics of Personal Exchange," *First Monday,* Vol. 5, No. 9 (September 2000).

29 "Privacy and Human Rights," p. 59.

30 "DoubleClick Turns Away from Ad Profiles," CNET News, January 8, 2002.

31 "Privacy Online: Fair Information Practices in the Electronic Marketplace," www.ftc.gov/reports/privacy2000/privacy2000.pdf.

32 "Survey: Internet Users Have More Control Over How Data Is Used," WashingtonPost.com, March 27, 2002.

33 "EPIC Alert," May 23, 2002, www.epic.org/alert/EPIC_Alert_9.10.html.

34 "Online Privacy Laws Debated Across US," Newsfactor.com, July 25, 2001, www.newsfactor.com/perl/story/6989.html.

35 http://europa.eu.int/comm/internal_market/en/dataprot/.

36 www.privcom.gc.ca/legislation/02_06_01_e.asp.

37 www.privacy.gov.au/act/index.html.

38 Garfinkel, p. 260.

39 Luba Vangelova, "Internet Libel Laws in Limbo," *Online Journalism Review,* September 26, 2002, www.ojr.org/ojr/law/1033079636.php.

40 Michael Overing, "Defining Jurisdiction on the Internet," *Online Journalism Review,* September 26, 2002, www.ojr.org/ojr/law/1033079518.php.

41 Rainie, p. 2.

42 Paraphrased from "Prepared Statement of the Federal Trade Commission on Privacy Online: Fair Information Practices in the Electronic Marketplace." Before the Committee on Commerce, Science, and Transportation, United States Senate, Washington, D.C., May 25, 2000.

43 Chris Hoofnagle, EPIC, message posted to New School University online ethics course, November 19, 2001, and Jason Catlett, Junkbusters.

44 Yair Galil, "Industry Wants to Opt-out of Opt-in," Internet Law Journal, April 16, 2001, http://tilj.com/content/ecomarticle04140102.htm.

45 Richard L. Johannesen, *Ethics in Human Communication*, 5th ed. (Prospect Heights, IL: Waveland, 2002), pp. 13–14.

46 Michelle Delio, "Yahoo's 'Opt-out' Angers Users," Wired News, April 2, 2002, www.wired.com/news/privacy/0,1848,51461,00.html.

47 Delio.

48 Robert Lemons, "Can You Trust TRUSTe?" ZDNet News, November 2, 1999, http://zdnet.com.com/2100–11–516377.html?legacy=zdnn.

49 Paul Boutin, "Just How Trusty is TRUSTe?" Wired News, April 9, 2002, www.wired.com/news/exec/0,1370,51624,00.html.

50 Boutin.

51 Boutin. This quote is from Seth Ross, an official with PC Guardian, a California company that makes computer security products.

52 Don Goldhamer, "Defining Privacy: A Privacy Primer," at http://home.uchicago.edu/~dhgo/privacy-intro/dsld02.html.

53 Larry Pryor and Paul Grabowicz, "Privacy Disclosure on News Sites Low," Online Journalism Review, June 13, 2001, www.ojr.org/ojr/ethics/1017956628.php.

54 Larry Pryor and Paul Grabowicz, "Who Gives a Damn About Privacy," Online Journalism Review, June 13, 2001, www.ojr.org/ojr/ethics/1017956617.php.

55 Center for Media Education, "TeenSites.com: A Field Guide to the New Digital Landscape," 2001, www.cme.org.

56 Center for Media Education, p. 36.

57 Center for Media Education, p. 18.

58 eBay, "Appendix to the Privacy Policy," http://pages.ebay.com/help/community/privacy-appendix2.html.

59 Rainie, p. 3.

60 California Healthcare Foundation and Internet Healthcare Coalition, "Ethics Survey of Consumer Attitudes About Health Web Sites." Study released September 2000. Available online at www.ihealthcoalition.org/content/CHCF-EthicsSurvey.pdf.

61 Clinton Wilder and John Soat, "The Ethics of Data," Information Week online, May 14, 2001, www.informationweek.com/837/dataethics.htm.

62 Internet Healthcare Coalition, "eHealth Code of Ethics," introduced May 24, 2000 (original emphasis). Available online at www.ihealthcoalition.org/ethics/ehcode.html.

63 Dana Hawkins, "Gospel of Privacy Guru: Be Wary Assume the Worst." Interview with Larry Ponemon, *US News and World Report* online, June 25, 2001,
www.usnews.com/usnews/nycu/tech/articles/010625/tech/privacy.htm.

64 Phone interview with attorney Michael Overing, July 15, 2002. Overing writes a legal column for Online Journalism Review.

65 Rosen, pp. 8–9.

66 Berners-Lee, p. 146.

67 Privacy Foundation, "The Extent of Systematic Monitoring of Employee e-mail and Internet Use," July 9, 2001. Available online at www.privacy-foundation.org/workplace/technology/extentpf.html.

68 Computer Professionals for Social Responsibility, "Electronic Privacy Principles." Available online at www.cpsr.org/program/privacy/privacy8 .htm.

69 Lisa Guernsey, "On the Job, the Boss Can Watch Your Every Online Move, and You Have Few Defenses," *New York Times*, December 16, 1999.

70 Computer Professionals for Social Responsibility.

71 SPJ code.

72 Susan Hornig Priest, *Doing Media Research* (Thousand Oaks, CA: Sage, 1996), p. 29.

73 Telephone interview with Jennifer Egan, June 7, 2002.

74 J.D. Lasica, "A Scorecard for Net News Ethics," Online Journalism Review, September 20, 2001, www.ojr.org.

75 E-mail interview, May 29, 2002.

76 Telephone interview, April 9, 2002.

77 E-mail interview, April 22, 2002.

78 John Schwartz, "Scouring the Internet in Search of the Tracks of Terrorists," *New York Times* on the Web, September 17, 2001, www.nytimes .com/2001/09/17/technology/17CRYP.html.

79 American Library Association, "Privacy: An Interpretation of the Library Bill of Rights." Statement adopted June 19, 2002. Available online at www.ala.org/alaorg/oif/privacyinterpretation.html. Laura Flanders, "Librarians Under Siege," *The Nation*, August 5, 2002. Accessed on Alternet.org, www.alternet.org/story.html?StoryID=13676. "A Chill in the Library." Editorial, *The St. Petersburg Times*, July 23, 2002, www.sptimes.com/2002/07/23/Opinion/A_chill_in_the_librar.shtml.

Chapter 5

1 *Palko v. Connecticut*, 302 U.S. 319, 325–27 (1937).

2 Alexander Meikeljohn, *Free Speech and Its Relation to Self Government* (New York: Harper, 1948), p. 27.

3 John Stuart Mill, *On Liberty*, *Great Books of the Western World*, edited by R. Hutchins (Chicago: Encyclopaedia Brittanica, 1952), pp. 274–293.

4 United Nations, Universal Declaration of Human Rights. Available online at www.un.org/Overview/rights.html.

5 United Nations, International Covenant on Civil and Political Rights, adopted and opened for signature, ratification and accession by General

Assembly resolution 2200A (XXI) of 16 December 1966 *entry into force* 23 March 1976, in accordance with Article 49. Available online at www.unhchr.ch/html/menu3/b/a_ccpr.htm.

6 From the South African Bill of Rights, Constitution of the Republic of South Africa, Act 108 of 1996. Available online at www.concourt.gov.za/constitution/const02.html.

7 Declan McCullagh, "Google Excluding Controversial Sites," CNET News, October 23, 2002, http://news.com.com/2100–1023–963132.html.

8 Robert W. McChesney, "The New Theology of the First Amendment: Class Privilege Over Democracy," *Monthly Review*, Vol. 49, No. 10 (1998), p. 17.

9 McChesney, p. 30.

10 Quoted in Thomas L. Tedford, *Free Speech in the United States*, 3ʳᵈ ed. (State College: Strata, 1997), p. 31.

11 Tedford, p. 27.

12 *New York Times v. Sullivan*, 376 U.S. 254 (1964).

13 See, for example, *Schenk v. United States*, 249 U.S. 47 (1919) and *Abrams v. United States*, 250 U.S. 616 (1919).

14 *Unites States v. Schwimmer*, 279 U.S. 644, 655 (1929). Justice Holmes dissenting.

15 *Gitlow v. New York*, 268 U.S. 652 (1925).

16 See *Robins v. Pruneyard Shopping Center,* 592 P.2d 341 (1979).

17 *Miller v. California*, 413 U.S. 15 (1973).

18 See *New York v. Ferber*, 458 U.S. 747 (1982), and *Osborne v. Ohio,* 495 U.S. 103 (1990).

19 *Brandenburg v. Ohio*, 395 U.S. 444 (1969).

20 See *Cohen v. California*, 403 U.S. 15 (1971), and *Gooding v. Wilson*, 405 U.S. 518 (1972).

21 Robert S. Peck, *Libraries, the First Amendment and Cyberspace* (Chicago: American Library Association, 2000), p. 31.

22 376 U.S. 254 (1964).

23 418 U.S. 323 (1974).

24 *United States v. Grace*, 461 U.S. 171, 177 (1983).

25 Peck, p. 35.

26 *Associated Press v. U.S.*, 326 U.S. 1, 20 (1945).

27 Tedford, p. 362.

28 *FCC v. Pacifica Foundation*, 438 U.S. 726 (1978).

29 *ACLU v. Reno*, 929 F. Supp. 824, 883 (E.D. Pa. 1996) (Dalzell, J.), *aff'd*, 117 S.Ct. 2329 (1997).

30 *Reno v. ACLU*, 96 U.S. 511 (1997).

31 See *ACLU et al. v. Ashcroft* No. 99-1324, http://www.ca3.uscourts.gov/opinarch/991324.pdf. The ACLU also has extensive case information on

its Web site, www.aclu.org. Other sites featuring coverage of Internet-related obscenity issues are Wired News (www.wired.com) and CNET News (news.com).

32 *Ashcroft v. Free Speech Coalition*, 00–795 (2002).

33 *American Library Association v. United States*, 01–1303 (2002).

34 Reuters, "Iran Reformers Use Net to Fight Press Ban," August 5, 2002. Accessed from CNET News, http://news.com.com/2100–1023–948403.html.

35 Lisa Bowman, "Enforcing Laws in a Borderless Web," CNET News, May 19, 2002, http://news.com.com/2100–1023–927316.html.

36 GILC, "Regardless of Frontiers: Protecting the Human Right to Freedom of Expression on the Internet," www.cdt.org/gilc/report.html, accessed October 22, 2002.

37 Bowman.

38 McCullagh.

39 Luba Vangelova, "Internet Libel Laws in Limbo," Online Journalism Review, September 26, 2002, www.ojr.org/ojr/law/1033079636.php.

40 Michael Overing, "Defining Jurisdiction on the Internet," Online Journalism Review, September 26, 2002, www.ojr.org/ojr/law/1033079518.php.

41 Bowman.

42 James W. Carey, *Communication as Culture* (New York: Routledge, 1989), p. 82.

43 SPJ Code of Ethics. Online at https://www.spj.org/ethics_code.asp.

44 Richard L. Johannesen, *Ethics in Human Communication*, 4th ed. (Prospect Heights, IL: Waveland Press, 1996), p. 256.

45 Haig Bosmajian, *The Language of Oppression* (Washington, DC: Public Affairs Press, 1974), pp. 1–10.

46 See Independent Media Center (Indymedia) newswire publishing policy, www.indymedia.org/fish.php3?file=www.indymedia.newswire, accessed October 25, 2002.

47 Internet Freedom, "The Hate Speech Controversy," Internet Freedom Web site, www.netfreedom.org/racism/position.html, accessed October 20, 2002.

48 MSNBC.com, "Code of Conduct," www.msnbc.com/news/556864.asp, accessed November 3, 2002.

49 Yahoo, "Terms of Service," http://docs.yahoo.com/info/terms/, accessed November 3, 2002.

50 Tom Spring, "Digital Hate Speech Roars," PCWorld.com, September 21, 2001, www.pcworld.com/news/article/0,aid,63225,00.asp.

51 Spring.

52 Ariana Eunjung Cha, "Online Companies Draw Fire for Removing 'Offensive' Postings," *Washington Post*, November 18, 2001, p. H1.

53 Cha.

54 Howard Zinn, *Declarations of Independence: Cross Examining American Ideology* (New York: Harper Perennial, 1990), p. 191.

55 Barb Palser, "Charting New Terrain," *American Journalism Review*, November 1999, pp. 20–31.

56 Barb Palser, "Too Hot for the Web," *American Journalism Review*, October 2000, http://216.167.28.193/article.asp?id=189.

57 Steven Mindich, "Thoughts on Political Pornography," *Boston Phoenix* Web site, June 4, 2002, www.bostonphoenix.com/pages/boston/daniel_pearl.html.

58 Mark Jurkowitz, "Phoenix, Web Firm Boost Access to Video of Reporter's Murder," *Boston Globe*, June 4, 2002, p. E1.

59 Bob Steele, "Pearl Photo: Too Harmful," Poynter Institute online column, June 7, 2002, www.poynter.org/talkaboutethics/.

60 Robert Spears, message posted to Online News listserv maintained by the Poynter Institute for Media Studies, June 6, 2002. Archived at http://talk.poynter.org/cgi-bin/lyris.pl?enter=online-news&text_mode=0&lang=english.

61 Nuremberg Files Web site, www.christiangallery.com/atrocity/.

62 *Planned Parenthood v. American Coalition of Life Activists,* No. 9935320p (July 10, 2002).

63 E-mail interview, March 25, 2002.

64 Taken from "Consequences of Computing: A Framework for Teaching," by Impact Computer Science Group, quoted in Janna Quitney Anderson and David Arant, "Ethics for Digital Information Providers," June 11, 2000, http://www.elon.edu/andersj/ethics.html.

65 Telephone interview, April 11, 2002.

66 Palser, "Charting New Terrain."

67 Palser, "Charting New Terrain."

68 Dianne Lynch, "Without a Rulebook," *American Journalism Review*, Vol. 20, No. 1 (January 1998), p. 40.

69 J.D. Lasica, "A Scorecard for Net News Ethics," Online Journalism Review, September 20, 2001, www.ojr.org.

70 Phil Nesbitt, "Tragedy in Photos: A New Standard?" American Press Institute Web site, September 14, 2001, http://americanpressinstitute.org/.

71 Message posted to Online News listserv maintained by the Poynter Institute for Media Studies, September 14, 2001. Archived at http://talk.poynter.org/cgi-bin/lyris.pl?enter=online-news&text_mode=0&lang=english.

72 Steve Outing, "Why Web News Has Different Standards Than Print, TV News," Poynter Institute online column, September 20, 2001, www.poynter.org/Terrorism/steve5.htm.

73 Message posted to Online News listserv maintained by the Poynter Institute, September 16, 2001. Archived at http://talk.poynter.org/cgi-bin/lyris.pl?enter=online-news&text_mode=0&lang=english.

74 Larry Gross, John Stuart Katz, and Jay Ruby, eds., *Image Ethics: The Moral Rights of Subjects in Photographs, Film, and Television* (New York: Oxford University Press, 1988), pp. v, 32.

75 Jurkowitz.

76 Declan McCullagh, "Besieged ISP Restores Pearl Vid," Wired News, May 28, 2002, www.wired.com/news/politics/0,1283,52818,00.html.

77 John Giuffo, "Linking and the Law," *Columbia Journalism Review*, September/October 2001, p. 11.

78 *Universal Studios et al. v. Reimerdes et al.*, August 17, 2000.

79 Giuffo.

Chapter 6

1 *Whole Earth Review*, May 1985, p. 49. Quotation used from a speech given by Brand in 1984.

2 Stewart Brand, *The Media Lab: Inventing the Future at MIT* (New York: Viking, 1987), p. 202. For readers interested in a full discussion of the slogan and how it has been used over time, we recommend Roger Clarke's Information Wants to Be Free Web page, available at www.anu.edu.au/people/Roger.Clarke/II/IWtbF.html. Clark is a Visiting Fellow in the Department of Computer Science at Australia's National University.

3 See, for example, John Perry Barlow, "The Economy of Ideas," Wired 2.03, March 1994. Retrieved from the World Wide Web: www.wired.com/wired/archive /2.03/economy.ideas.html?pg=1.

4 Christopher D. Hunter, "Copyright and Culture." Paper presented at the National Communication Association Convention (Seattle), November 2000, www.asc.upenn.edu/usr/chunter/copyright_and_culture.html.

5 Ron Bettig, *Copyrighting Culture: The Political Economy of Intellectual Property* (Boulder, CO: Westview Press, 1996), chap. 1.

6 Bettig, p. 18.

7 Hunter.

8 See Wikipedia: The Free Encylopedia Web site at http://www.wikipedia.org/wiki/Statute_of_Anne.

9 Michael C. McFarland, SJ (Society of Jesus, i.e., Jesuits), "Intellectual Property, Information, and the Common Good." Paper presented at the Fourth Annual Ethics and Technology Conference, Boston College, Boston, MA, June 5, 1999. Available online at http://infoeagle.bc.edu/bc_org/avp/law/st_org/iptf/commentary/index.html.

10 Lawrence Lessig, *The Future of Ideas: The Fate of the Commons in a Connected World* (New York: Random House, 2001), p. 106.

11 U.S. Copyright Office, "A Brief History and Overview," www.copyright.gov/docs/circ1a.html.

12 Jessica Litman, *Digital Copyright* (Amherst, NY: Prometheus, 2001), pp. 22–69.

13 Lessig.

14 U.S. Copyright Office, "Copyright Basics," updated September 16, 2002, www.copyright.gov/circs/circ1.html#wci. Accessed November 3, 2002.

15 U.S. Code, Title 17, Chapter 1, Section 106. Online at http://www4.law.cornell.edu/uscode/17/106.html.

16 U.S. Copyright Office, "Copyright Basics."

17 U.S. Code, Title 17, Chapter 1, Section 107. Online at http://www4.law.cornell.edu/uscode/17/107.html.

18 *Folsom v. Marsh,* 9 F. Cas. 342, 348 (No. 4, 901) (CCD Mass. 1841).

19 U.S. Code, Title 17, Chapter 1, Section 107.

20 U.S. Code, Title 17, Chapter 1, Section 107.

21 Oberlander is counsel for *Forbes* magazine and a member of the Media Studies faculty at the New School University.

22 *Campbell v. Acuff-Rose Music, Inc.,* 510 U.S. 569 (1994).

23 Jack Valenti, "A Clear Present and Future Danger." Testimony to the Senate Committee on Foreign Relations, February 12, 2002. Available online at the MPAA Web site, www.mpaa.org/jack/2002/2002_02_12b.htm.

24 Litman, pp. 26–27.

25 *MAI Systems Corp. v. Peak Computer, Inc.* 991 F.2d 511 (9th Cir. 1993).

26 *Apple Computer v. Formula International,* 594 F. Supp. 617.

27 Litman, pp. 27–28.

28 *Eldred et al. v. Ashcroft* 537 U.S. (2003) http://www.supremecourtus.gov/opinions/02pdf/01-618.pdf. For more on Eldred's Web site and his legal challenge see Lessig, pp. 122–123, 196–199, and Andrea L. Foster, "A Bookworm's Battle," *The Chronicle of Higher Education,* Vol. 49, No. 9 (October 25, 2002), p. A35.

29 The economists' amicus brief is online at http://eon.law.harvard.edu/openlaw/eldredvashcroft/supct/amici/economists.pdf.

30 Doug Isenberg, "The 'Other' Digital Millennium Act," Gigalaw.com, October 2001, www.gigalaw.com/articles/2001-all/isenberg-2001–10-all.html.

31 Isenberg.

32 Gerald R. Ferrera, Stephen D. Lichtenstein, Margo E.K. Reder, Ray August, and William T. Schiano, *Cyberlaw: Your Rights in Cyberspace* (Canada: Thompson Learning, 2001), pp. 59–60.

33 Declan McCullagh, "What Hollings' Bill Would Do," Wired News, March 22, 2002, www.wired.com/news/politics/0,1283,51275,00.html, and Digital Consumer at www.digitalconsumer.org.

34 *Feist v. Rural Telephone Service Co.,* 499 U.S. 340 (1991).

35 Yochai Benkler, "The Battle Over the Institutional Ecosystem in the Digital Environment," *Communications of the ACM,* Vol. 44, No. 2 (February 2001), p. 86.

36 Kim Campbell, "Fees Threaten to Silence Web Radio," *Christian Science Monitor*, April 4, 2002, and F. Timothy Martin, "Jesse Helms, Web Radio's Hero," Salon.com, November 19, 2002, www.salon.com/tech/feature/2002/11/19/helms_web_radio/.

37 Shelly Warwick, "Is Copyright Ethical? An Examination of the Theories, Laws, and Practices Regarding the Private Ownership of Intellectual Work in the United States." Paper presented at the Intellectual Property and Technology Forum, June 5, 1999, Boston College, Boston, MA.

38 Lessig; Benkler; and "From Consumers to Users: Shifting the Deeper Structures of Regulation Toward Sustainable Commons and User Access," *Federal Communications Law Journal*, Vol. 52, No. 3, pp. 561–579.

39 Valenti.

40 George Johnson, "If You Can't Touch It, Can You Steal It?" *New York Times* on the Web, December 16, 2001, www.nytimes.com/2001/12/16/weeinreview/16JOHN.html.

41 Edwin C. Hettinger, "Justifying Intellectual Property Rights," *Philosophy and Public Affairs*, Vol. 18, No. 1 (Winter 1989), pp. 31–52.

42 Open Democracy, "Let's Share!" Interview with Richard Stallman, May 30, 2002. Open Democracy Web site, www.opendemocracy.net.

43 Open Democracy.

44 Bernhard Warner, "Copy-Proof CDs Cracked with 99-cent Marker Pen," Reuters, May 20, 2002. Article accessed through Yahoo! News, May 25, 2002.

45 Warner.

46 *Los Angeles Times v. Free Republic*, case no. 98–7840. Judge's tentative ruling on the fair use issue is available online at http://techlawjournal.com/courts/freerep/19991108.htm).

47 C. Edwin Baker, "Market Threats to Press Freedom." Address to the 15th Annual Symposium on American Values, Angelo State University, San Angelo, TX, October 12, 1998. Online at www.angelo.edu/events/university_symposium/1998/Baker.htm.

48 Ferrera et al., p. 41.

49 Barb Palser, "Charting New Terrain," *American Journalism Review*, November 1999, pp. 20–31.

50 Palser, "Charting New Terrain."

51 Tim Berners-Lee, *Weaving the Web: The Original Design and Ultimate Destiny of the World Wide Web by Its Inventor* (San Francisco: Harper Collins, 1999), p. 140.

52 *Ticketmaster Corp., et al. v. Tickets.Com, Inc.,* U.S. Dist. Lexis 12987 (C.D. Ca., August 10, 2000).

53 Richard Spinello, "Web Site Linking: Right or Privilege." Paper presented at the Fourth Annual Ethics and Technology Conference, Boston, MA, June 5, 1999.

54 Berners-Lee, pp. 1–6.

55 Comment by Jamie Kellner, an executive with AOL-Time Warner, quoted in Staci Kramer, "Content's King," *Cableworld*, April 29, 2002.

56 Michelle Finley, "Attention Editors: Deep Link Away," Wired News, March 30, 2000, online www.wired.com/news/politics/0,1283,35306,00 .html.

57 www.dontlink.com/.

58 Wayne Robins, "Content and Its Discontents," *Editor & Publisher*, Vol. 133, No. 49 (December 4, 2000), p. 26.

59 Valenti.

60 Valenti.

61 McFarland.

62 Litman, chap. 10.

63 John Perry Barlow, "The Next Economy of Ideas," Wired, October 2000, online at www.wired.com/wired/archive/8.10/download.html.

64 U.S. District Judge Harry Hupp in *Ticketmaster Corp., et al. v. Tickets.Com, Inc.*, 2000 WL 1887522 (C.D. Cal.).

65 Michael Overing, "Impermissible Links," Online Journalism Review, May 23, 2001, http://www.ojr.org/ojr/law/p1017958882.php

66 NPR.org.

67 Jeffrey A. Dvorkin, "Linking to the Web site: Irate Bloggers and Other New Ideas," NPR.org, June 28, 2002, http://www.npr.org/yourturn/ombudsman/index.html.

68 Dvorkin.

69 Staci Kramer, "NPR's Mixed Messages," Online Journalism Review, June 25, 2002, http://www.ojr.org/ojr/kramer/p1025044888.php.

70 E-mail message posted June 24, 2002, to Online News discussion list maintained by the Poynter Institute for Media Studies. Lists are archived at http://www.poynter.org.

71 E-mail message posted June 25, 2002, to Online News discussion list maintained by Poynter Institute for Media Studies. Archived at www .poynter.org.

72 Dvorkin.

73 Kramer.

74 Posted on Doctorow's Weblog Boing Boing: A Directory of Wonderful Things, June 27, 2002, http://boingboing.net/2002_06_01_archive.

Part 3 Introduction

1 Deuze quoted in Mark Deuze and Daphna Yeshua, "Online Journalists Face New Ethical Dilemmas: Lessons from the Netherlands," *Journal of Mass Media Ethics*, Fall 2001, p. 275.

2 Posted on September 18, 1993, via an e-mail discussion list, now defunct; quoted in Robert Berkman, "Online Journalists Line Up Behind Online," *Montana Journalism Review*, Fall 1994, p. 43.

3 Howard Kurtz, "SpinMaster," *Washington Post* online, March 8, 1998, p. W11.
4 "The Drudge Phenomenon," www.nyu.edu/gsas/dept/journal/Drudge/copy/drudge_phenom.htm.
5 Retrieved from the World Wide Web: http://update.usatoday.com/go/newswatch/97/april/nw0420–2.htm.
6 See www.onlinenewsassociation.org/AboutONA/Mission.htm.
7 See www.editorandpublisher.com/editorandpublisher/business_resources/mediastats.jsp.
8 See www.pewinternet.org/reports/toc.asp?Report=80. See part 6 of the study.
9 See www.cyberjournalist.net/features/netratings/1202netratings.htm.
10 See http://newslink.org/topsites.html.

Chapter 7

1 Michael Singletary and Richard Lipsky, "Accuracy in Local TV News," *Journalism Quarterly*, Summer 1977, pp. 363–364.
2 Ron F. Smith, *Groping for Ethics in Journalism*, 4th ed. (Ames: Iowa State Press, 1999), pp. 56–67.
3 See, for example, Bill Kovach and Tom Rosenstiel, *Warp Speed: America in the Age of Mixed Media Culture* (Washington, DC: Century Foundation Press, 1999), p. 56, and Smith, pp. 62–63.
4 See Kovach & Rosenstiel, p. 56.
5 The Digital Journalism Credibility Study found that media professionals were more prone to being critical of online news sites' credibility than was the general public. This gap in perception prompted the study's publishers to raise the question: "Is there something the media perceives or knows about the ethics and practices of online news organizations or operations that the public does not know? Or are traditional media just being resistant to online news?" (See www.onlinenewsassociation.org/programs/credibility-study.pdf.)

   Another study of user's perceptions of the credibility of various media was presented by the University of Miami's School of Communication in a paper presented at the Association for Education in Journalism and Mass Communication (AEJMC) 2002 conference. In an e-mail interview, Rasha Abdulla said that: "The study showed that trustworthiness and timeliness are factors that online news readers value the most about their online experience. The 'trustworthiness' factor included items such as 'trustworthy, believable, and accurate'." (Rasha A. Abdulla, Bruce Garrison, Michael Salwen, Paul Driscoll, and Denise Casey, "The Credibility of Newspapers, Television News and Online News." Retrieved from the World Wide Web: www.miami.edu/com/car/miamibeach1.htm.)

6　Posting to the online-news listServ (online-news@poynter.org), December 23, 2002.

7　E-mail correspondence, December 24, 2002.

8　Posting to the online-news listServ (online-news@poynter.org), December 25, 2002.

9　Posting to the online-news listServ (online-news@poynter.org), December 25, 2002.

10　See www.journalists.org/Programs/Research.htm.

11　Janna Quitney Anderson and M. David Arant, "Newspaper Online Editors Support Traditional Standards," *Newspaper Research Journal,* Fall 2001.

12　Digital Journalism Credibility, p. 37.

13　E-mail correspondence, December 24, 2002.

14　Digital Journalism Credibility, p. 38.

15　Digital Journalism Credibility, p. 36.

16　Digital Journalism Credibility, pp. 34–39.

17　E-mail correspondence, December 24, 2002.

18　Dure's report, which was completed in 2000, can be found on his personal Web site at http://hometown.aol.com/bdure/project/index.html.

19　Mark Deuze, "Online Journalists Face New Ethical Dilemmas: Lessons from the Netherlands," *Journal of Mass Media Ethics,* Fall 2001, pp. 273–292.

20　Deuze, p. 279. Deuze makes available follow-up studies to his research on his Web site at users.fmg.uva.nl/mdeuze/bibl.htm.

21　Frank Sennett, *Editor & Publisher Online,* February 8, 1999.

22　See, for example, "Online News Stories That Change Behind Your Back," SlashDot, May 9, 2002. Retrieved from the World Wide Web: http://slashdot.org/features/02/05/08/1924240.shtml?tid=149, which discussed the changes made by CNN/Money on its article online about the Microsoft antitrust trial.

Chapter 8

1　Society of Professional Journalists Web page, http://www.spj.org/spj_about.asp.

2　Leonard Downie Jr. and Robert Kaiser, *The News About the News: American Journalism in Peril* (New York: Knopf, 2002).

3　For more details, see www.middleberg.com/toolsforsuccess/fulloverview.cfm.

4　See www.pewinternet.org/reports/index.asp, from "Summary of Findings."

5　www.middleberg.com/toolsforsuccess/fulloverview.cfm.

6 www.inms.umn.edu/convenings/newslibraryincrisis/main.htm.
7 Excerpted, with permission, from *The Skeptical Searcher: The Information Advisor's Guide to Evaluating Business Web Data, Sites and Sources* (Medford, NJ: Information Today, 2003).
8 *Columbia Journalism Review,* Vol. 35, No. 4, pp. 36–38 (November-December 1996).
9 Paraphrased from a participant's Web page at http://www.content-exchange.com/cx/html/newsletter/1–21/news1–21.5.htm. (Official page at Poynter.org describing the event is called "Finding a Niche for Investigative Journalism," http://209.241.184.41/centerpiece/021000-index.htm.
10 Ann M. Brill, "Online Journalists Embrace New Marketing Function," *Newspaper Research Journal,* Vol. 22, No. 2 (Spring 2001), p. 28.
11 David H. Weaver and G. Cleveland Wilhoit, *The American Journalist in the 1990s: U.S. News People at the End of an Era* (Bloomington: University of Indiana Press, 1986).
12 Using search engines and news search engines to uncover previously published articles also has significant limitations. Online news sites vary widely in how much of their archive is freely available on their Web site. A few make everything available and at no charge; others make only the last few days or weeks free and charge a fee to users who want older news items. Some don't have any archives available at all on their Web site, free or fee. (You can view a very useful table created by the News Division of the Special Libraries Association that lists the archiving policies for U.S. and international newspapers. See www.ibiblio.org/slanews/internet/archives.html.) Some news sites require passwords or registration to access their archives— and in those cases, these news stories are part of what is called "The Invisible Web"— that is, information that resides on the Web but cannot be indexed by search engines (for more about the invisible Web, see Chris Sherman and Gary Price, *The Invisible Web* [Medford, NJ: CyberAge Books, 2001]).

So although you can now search for news stories from more publishers, and from around the world, and do so easier and cheaper than ever before, you are not searching "everything." In fact, you are more than likely missing the deeper news archives that can only be researched by paying to search the archives of an individual news source's site, via a fee-based database or CD-ROM or by checking a library physical collection in print or microfilm.
13 E-mail correspondence, December 27, 2002.
14 "Study: Media Plays Fast, Loose with Ethics Online," by Victor Merina, Poynter Fellow, March 6, 2000, http://legacy.poynter.org/centerpiece/030600-index.htm.
15 See Sherman & Price.
16 www.middleberg.com/toolsforsuccess/fulloverview.cfm.

Chapter 9

1 William Leiss, Stephen Kline, and Sut Jhally, "Advertising, Consumers, and Culture." In David Crowley and Paul Heyer (Eds.), *Communication in History* (New York: Longman, 1999), p. 210, and Ron F. Smith, *Groping for Ethics in Journalism,* 4th ed. (Ames: Iowa State Press, 1999), p. 289.

2 Ben Bagdikian, *The Media Monopoly,* 5th ed. (Boston: Beacon Press, 1997), p. 135.

3 Thomas Leonard, "Lessons from L.A.: The Wall: A Long History," *Columbia Journalism Review*, January/February 2000.

4 Leonard.

5 American Society of Newspaper Editors, "Statement of Principles," online at www.asne.org/kiosk/archive/principl.htm. The original "Canons of Journalism" have been posted on numerous sites, including www.iit.edu/departments/csep/PublicWWW/codes/coe/Sigma_Delta_Chi_New.html.

6 SPJ Code of Ethics, online at http://www.spj.org/ethics_code.asp.

7 RTNDA Code of Ethics, online at www.rtnda.org/ethics/coe.shtml.

8 Federal Communications Act of 1934, Sec. 317. (a) (1).

9 Federal Communications Act of 1934, Sec. 73.119 (f).

10 George Seldes, *Lords of the Press* (New York: Blue Ribbon Books, 1941); see also Lawrence Soley, *Censorship Inc.: The Corporate Threat to Free Speech in the United States* (New York: Monthly Review Press, 2002), chap. 8 for a summary of Seldes' reporting on advertisers.

11 Quoted in Bagdikian, pp. 156–157.

12 Quoted in Bagdikian, p. 157.

13 Bagdikian, pp. 168–173.

14 Bagdikian, p. 15.

15 Lawrence Soley, "The Power of the Press Has a Price," *Extra!*, July/August 1997.

16 "The Truth About Self-Censorship," *Columbia Journalism Review*, May/June 2000.

17 Janine Jackson and Peter Hart, "Fear and Favor 2000: How Power Shapes the News," *Extra!*, May/June 2001, pp. 15–22; for similar accounts also see "Fear and Favor 2001," *Extra!*, March/April 2002, pp. 20–27.

18 "Gambling with the Future," special supplement to *Columbia Journalism Review*, November/December 2001.

19 "On the Road to Irrelevance," special supplement to *Columbia Journalism Review*, November/December 2002.

20 Smith, p. 270.

21 Louis Chunovic, "Advertisers Notice Younger Viewers," *Electronic Media* online, December 31, 2001, www.emonline.com/advertise/123101advertisers.html. Accessed January 12, 2002.

22 "Reader Profile," *New York Times* on the Web, January 16, 2003, www.nytimes.com/adinfo/audience_profile.html.

23 C. Edwin Baker, "Market Threats to Press Freedom." Speech delivered to the Symposium on American Values at Angelo State University, San Angelo, TX, October 12, 1998, online at www.angelo.edu/events/university _symposium/1998/baker.htm; for more of Baker's work on advertising see his book *Advertising and a Democratic Press* (Princeton, NJ: Princeton University Press, 1994); Bagdikian also quotes similar remarks from Chandler in *The Media Monopoly,* 5th ed., p. 116.

24 Bagdikian, p. 110.

25 John H. McManus, *Market-Driven Journalism* (Thousand Oaks, CA: Sage, 1994), p. 197.

26 See "The National Public Survey on White Collar Crime." A summary of the study is online at www.nw3c.org/surveyresults.htm.

27 Robert Jensen, "The Sport of Business/The Business of Sport," *Media Ethics*, Vol. 11, No. 2 (Spring 2000), pp. 4, 20–21.

28 Bagdikian, p. 155.

29 Quoted in Lawrence Soley, *Censorship Inc: The Corporate Threat to Free Speech in the United States* (New York: Monthly Review Press, 2002), p. 229; see also Jim Naureckas, "Corporate Ownership Matters: The Case of NBC," *Extra!*, November/December 1995.

30 Quoted in Ben Bagdikian, *The Media Monopoly,* 6th ed. (Boston: Beacon Press, 2000), p. xxv; see also "A CEO's Point of View," on the ASNE Web site, which includes a transcript of Willes' address to the ASNE on April, 3, 1998, www.asne.org/kiosk/archive/convention/conv98/willes .htm.

31 Robert G. Magnuson, "How Do We Protect Integrity in an Increasingly Complex World?" Poynter.org, no date on the document. Accessed January 7, 2003, www.poynterextra.org/centerpiece/jbv/magnuson.htm.

32 Greg Mitchell, "Wall Comes Tumbling Down," *Editor & Publisher,* December 4, 1999. See also Online News Association, "Digital News Credibility Study," http://www.journalists.org/Programs/Research.htm.

33 Bagdikian, *The Media Monopoly,* 6th ed., p. xxvii.

34 Farai Chideya quoted in Dirk Smillie, "Panelists Say Separating Content, Commerce a Major Web Issue," Freedom Forum online, March 2, 2000, www.freedomforum.org/templates/document.asp?documentID=11780. Accessed December 15, 2001.

35 Online News Association, "Digital Journalism Credibility Study," January 31, 2002, online at www.onlinenewsassociation.org/Programs/ Research.htm.

36 ONA, p. 50.

37 "At Least 16 Die as Car Bomber Hits Israeli Bus," June 5, 2002. See www.nytimes.com/2002/06/05/international/05CND-MIDE.html.

38 Telephone interview with Aly Colón, April 9, 2002.

39 Bill Carter, "Skipping Ads? TV Gets Ready to Fight Back," *New York Times*, January 10, 2003, p. C-1.

40 NYTimes.com press release, December 6, 2001, http://www.businesswire
.com/nyt-digital/.

41 Barb Palser, "Advertorial with a Twist," *American Journalism Review*,
May 2002, http://216.167.28.193/article.asp?id=2510

42 ONA.

43 "Race for Ad Dollars Tests Online News," CBS MarketWatch.com, December 29, 2001.

44 E-mail interview with Christine Mohan, *New York Times* Digital
spokesperson, August 20–21, 2002.

45 "Online Publishers Struggle to Decide Where to Draw Line on Advertising," *Wall Street Journal* Online, January 29, 2002.

46 E-mail interview with Mohan.

47 Online News discussion list, December 14, 2001. Archived at http://
talk.pointer.org/cgi-bin/lyris.pl?visit=online-news&id=188862944.

48 Palser.

49 Palser.

50 E-mail interview with Mohan.

51 "NYTimes.com Finds Promotion of HBO Project Fit to Print," *Wall Street
Journal* Online, March 14, 2002.

52 The complete "Best Practices for Digital Media" document can be viewed
at http://asme.magazine.org/guidelines/new _media.html.

53 The complete Poynter/ASNE protocols can be viewed at www.poynter
.org/dj/Projects/newmediaethics/me samprot.htm#integrity. Also quoted
in David Arant and Janna Quitney Anderson, "Online Media Ethics: A
Survey of U.S. Daily Newspaper Editors," http://www.elon.edu/andersj/
onlinesurvey.html.

54 Howard Finberg, "Report: Online News Widely Accepted as Credible,"
Poynter.org, February 1, 2002. Accessed March 12, 2002, www.poynter
.org/content/content_view.asp?id=3509.

55 Finberg.

56 ONA, p. 34.

57 Barb Palser, "Charting New Terrain," *American Journalism Review*, November 1999.

58 CNET Networks, "Editorial and Disclosure Statement," online at www
.cnet.com/aboutcnet/0–7251795–7–7283134.html. Last updated October
10, 2000. Accessed December 23, 2002.

59 Magnuson.

60 Quoted in Staci Kramer, "Merrill Brown: After the 'Heyday'," Online Journalism Review, June 13, 2002, www.ojr.org/ojr/kramer/1024012308 .php.

61 A few news-oriented sites that accept little or no advertising include Alter-
net (www.alternet.org), Tom Paine (www.tompaine.com), Common
Dreams (www.commondreams.org), Znet (www.zmag.org), One World
Network (www.oneworld.net), and Inter Press News Service (ipsnews .net).

Afterword

1 See introductory material in the 5th ed. (Boston: Beacon Press, 1997), p. xlvi.

2 See 6th ed. (Boston: Beacon Press, 2000), p. viii.

3 See Robert W. McChesney, *Rich Media, Poor Democracy: Communications Politics in Dubious Times* (Champaign: University of Illinois Press, 1999).

4 See McChesney; also Douglas Dowd, *Capitalism and Its Economics* (London: Pluto Press, 2000).

5 Bagdikian, 5th ed., p. xv.

6 James Fallows, "Internet Illusions," *New York Review of Books*, November 16, 2000. Online at www.nybooks.com/articles/13891. Accessed October 14, 2002. For a more detailed view by a prominent cyberspace utopian, see Howard Rheingold, *The Virtual Community: Homesteading on the Electronic Frontier* (Reading, MA: Addison-Wesley, 1993).

7 Darrell West, *The Rise and Fall of the Media Establishment* (Boston: Bedford/St. Martin's, 2001); see in particular chaps. 6 and 7.

8 See Herbert I. Schiller, *Information Inequality* (New York: Routledge, 1996); McChesney, chap. 3; and David Resnick, "Politics on the Internet: The Normalization of Cyberspace." In Chris Toulouse and Timothy W. Luke (Eds.), *The Politics of Cyberspace* (New York: Routledge, 1998), pp. 48–68.

9 McChesney, p. 183.

10 McChesney, p. 119.

11 Quoted in Patricia Aufderheide, *Daily Planet: A Critic on the Capitalist Cultural Beat* (Minneapolis: University of Minnesota Press, 2000), pp. 185–186.

12 Quoted in Herbert I. Schiller, *Culture Inc: The Corporate Takeover of Public Expression* (New York: Oxford University Press, 1989), p. 122.

13 Aufderheide, p. 186.

14 George Albert Gladney and James D. Ivory, "Attitudes of Relational Engagement in Cyberspace: Uncovering Monologic Potential and Growth." Paper presented at the Union of Democratic Communications Annual Conference, October 11, 2002, State College, PA.

15 Michael Katz, "Look for 1-Billion Web Users by '05," *Media Life Magazine*, August 16, 2002, www.medialifemagazine.com/news2002/aug02/aug12/5_fri/news4friday.html. Accessed December 14, 2002.

16 Gladney & Ivory, p. 19.

17 See Schiller, *Information Inequality*, chap. 5.

18 Lawrence Lessig, *Code: And Other Laws of Cyberspace* (New York: Basic Books, 1999).

19 Fallows.

20 "FCC Classifies Cable Modem Service as 'Information Service'," FCC Press Release, March 14, 2002. See www.fcc.gov/Bureaus/Cable/News_Releases/2002/nrcb0201.html.

21 See *The Future of Ideas: The Fate of the Commons in a Connected World* (New York: Random House, 2001).

22 Center for Digital Democracy, "Canada's Canarie: Building Community Models for Broadband," November 27, 2002, www.democraticmedia.org/news/washingtonwatch/Canarie.html.

23 Center for Public Integrity, "Off the Record: What Media Companies Don't Tell You About Their Legislative Agendas" (Washington, DC: The Center for Public Integrity, 2000), www.publicintegrity.org/dtaweb/index.asp?L1=20&L2=21&L3=0&L4=0&L5=0&State=.

24 "The Death of the Internet?" TomPaine.com, October 24, 2002, www.tompaine.com/feature.cfm/ID/6600.

25 Norman Solomon, "Simulating Democracy Can Be A Virtual Breeze," "Media Beat" column, Fairness and Accuracy in Reporting online, May 24, 2001, www.fair.org/media-beat/010524.html. Accessed June 25, 2001.

26 *The Nature of Markets in the World Wide Web* (Palo Alto, CA: Xerox Palo Alto Research Center, 1999).

27 Bernardo Huberman, "Patterns in the World Wide Web," *The Library of Economics and Liberty*, August 26, 2002, www.econlib.org/library/Columns/Hubermanpatterns.html. Accessed December 13, 2002.

28 www.comscore.com/news/mm/pr_mm_101501.htm.

29 Amy Langfield, "Net News Lethargy," Online Journalism Review, April 3, 2002, www.ojr.org/ojr/reviews/1017864558.php cite?.

30 Robert W. McChesney, "Power to the Producers," *Boston Review*, Summer 2001.

31 McChesney, "Power to the Producers."

32 "Gotta Have Faith," *New York Times*, December 17, 2002, p. 35.

33 "TBO.com: Then and Now," Online Journalism Review, March 13, 2002, www.ojr.org/ojr/trends/p1016158090.php.

References

Books

Bagdikian, Ben H. *The Media Monopoly,* 5th ed. (Boston: Beacon Press, 1997).

Bagdikian, Ben H. *The Media Monopoly,* 6th ed. (Boston: Beacon Press, 2000).

Baker, C. Edwin. *Advertising and a Democratic Press* (Princeton, NJ: Princeton University Press, 1994).

Barnouw, Erik. *A History of Broadcasting in the United States,* 3 vols. (New York: Oxford University Press, 1966–1970).

Barnouw, Erik. *The Sponsor* (New York: Oxford University Press, 1979).

Berners-Lee, Tim. *Weaving the Web.* (San Francisco: Harper/San Francisco, 1999).

Bettig, Ronald V. *Copyrighting Culture: The Political Economy of Intellectual Property* (Boulder, CO: Westview Press, 1996).

Bosmajian, Haig. *The Language of Oppression* (Washington DC: Public Affairs Press, 1974).

Bourdieu, Pierre. *On Television* (New York: New Press, 1998).

Carey, James W. *Communication as Culture* (New York: Routledge, 1992), especially chaps. 3 and 6.

Crowley, David, and Paul Heyer (eds.). *Communication in History* (New York: Longman, 1999).

Dewey, John. *The Public and Its Problems* (Athens: University of Ohio Press, 1983). Originally published in 1927.

Downie, Leonard Jr., and Robert G. Kaiser. *The News about the News: American Journalism in Peril* (New York: Knopf, 2002).

Fallows, James. *Breaking the News: How the Media Undermine American Democracy* (New York: Vintage, 1997—paperback).

Ferrera, Gerald R., Stephen D. Lichtenstein, Margo E.K. Reder, Ray August, and William T. Schiano. *Cyberlaw: Your Rights in Cyberspace* (Toronto, Canada: Thompson Learning, 2001).

Garfinkel, Simson. *Database Nation: The Death of Privacy in the 21ˢᵗ Century* (Sebastopol, CA: O'Reilly, 2001—paperback).

Herman, Edward S., and Noam Chomsky. *Manufacturing Consent: The Political Economy of Mass Media* (New York: Pantheon, 1988). See also the 2002 edition with a new introduction, also by Pantheon.

Johannesen, Richard L.. *Ethics in Human Communication*, 5ᵗʰ ed. (Prospect Heights, IL: Waveland Press, 2002).

Johnson, Deborah G. *Computer Ethics,* 3ʳᵈ ed. (Upper Saddle River, NJ: Prentice Hall, 2001).

Kovach, Bill, and Tom Rosenstiel. *Warp Speed: America in the Age of Mixed Media Culture* (Washington DC: Century Foundation, 1999).

Kovach, Bill, and Tom Rosenstiel. *The Elements of Journalism: What Newspeople Should Know and the Public Should Expect* (New York: Crown, 2001).

Lessig, Lawrence. *The Future of Ideas: The Fate of the Commons in a Connected World* (New York: Random House, 2001).

Lippmann, Walter. *Public Opinion* (New York: Free Press, 1965). Originally published in 1922.

Litman, Jessica. *Digital Copyright* (Amherst, NY: Prometheus, 2001).

McChesney, Robert W. *Rich Media, Poor Democracy: Communications Politics in Dubious Times* (Champaign: University of Illinois Press, 1999). Paperback edition by New Press, New York City, 2000.

McManus, John H. *Market-Driven Journalism* (Thousand Oaks, CA: Sage, 1994).

Meyer, Philip. *Ethical Journalism* (New York: Longman, 1987).

Peck, Robert S. *Libraries, the First Amendment and Cyberspace* (Chicago: American Library Association, 2000).

Postman, Neil. *Amusing Ourselves to Death* (New York: Penguin, 1985).

Rosen, Jeffrey. *The Unwanted Gaze* (New York: Random House, 2000).

Scheuer, Jeffrey. *The Soundbite Society* (New York: Routledge, 2001).

Schiller, Herbert I. *Culture, Inc: The Corporate Takeover of Public Expression* (New York: Oxford University Press, 1989).

Schiller, Herbert I. *Information Inequality: The Deepening Social Crisis in America* (New York: Routledge, 1996).

Schudson, Michael. *Discovering the News: A Social History of American Newspaper* (New York: Basic Books, 1978).

Serrin, William, and Judith Serrin (eds.). *Muckraking: An Anthology of the Journalism that Changed America* (New York: New Press, 2002).

Smith, Ron F. *Groping for Ethics in Journalism*, 4ᵗʰ ed. (Ames: Iowa State Press, 1999).

Soley, Lawrence. *Censorship, Inc.: The Corporate Threat to Free Speech in the United States* (New York: Monthly Review Press, 2002).

Stevens, Mitchell. *A History of News: From the Drum to the Satellite* (New York: Viking, 1988).

Tedford, Thomas L., and Dale A. Herbeck. *Freedom of Speech in the United States,* 4ᵗʰ ed. (State College, PA: Strata, 2001).

West, Darrel. *The Rise and Fall of the Media Establishment* (Boston: Bedford/ St. Martins, 2001).

Zinn, Howard. *Declarations of Independence: Cross Examining American Ideology* (New York: Harper Perennial, 1990).

Periodicals, Web Journals and Academic Journals

American Journalism Review (www.ajr.org).

Broadcasting and Cable (www.broadcastingandcable.com).

Columbia Journalism Review (www.cjr.org).

CyberJournalist (http://cyberjournalist.net/).

Editor & Publisher (www.editorandpublisher.com).

Electronic Media (www.emonline.com),

Extra! (www.fair.org/extra/).

First Monday (www.firstmonday.dk/).

Gigalaw (www.gigalaw.com/).

Journal of Computer Mediated Communication (www.ascusc.org/jcmc/).

Journal of Mass Media Ethics.

Journalism and Mass Communication Quarterly.

Journalism History.

Journalism Studies.

Media Ethics.

Online Journalism Review (www.ojr.org).

Quill.

St. Louis Journalism Review (www.stljr.org/).

Television and New Media.

Organizations on the Web

American Civil Liberties Union "Cyber-Liberties" page (www.aclu.org/ Cyber-Liberties/Cyber-Libertiesmain.cfm).

American Library Association (www.ala.org/).

American Press Institute (www.americanpressinstitute.org/).

American Society of Newspaper Editors (www.asne.org).

Association for Education in Mass Communication and Journalism (http:// aejmc.org).

Berkman Center for Internet and Society (http://cyber.law.harvard.edu/).

Center for Democracy and Technology (www.cdt.org/).

Center for Digital Democracy (www.democraticmedia.org).

Committee to Protect Journalists (www.computerpress.org).

Computer Professionals for Social Responsibility (www.cpsr.org).

Consumer WebWatch (www.consumerwebwatch.org/).

Creative Commons (www.creativecommons.org/).

Electronic Frontier Foundation (www.eff.org).

Electronic Privacy Information Center (www.epic.org).

Fairness and Accuracy in Reporting (www.fair.org).

Federal Communications Commission (www.fcc.gov).

Federal Trade Commission (www.ftc.gov).

Freedom Forum (www.freedomforum.org).

Index on Censorship (www.indexonline.org/).

Investigative Reporters and Editors (www.ire.org).

Junkbusters (www.junkbusters.com/).

National Association of Black Journalists (www.nabj.org).

National Association of Hispanic Journalists (www.nahj.org).

National Institute for Computer Assisted Reporting (www.nicar.org).

National Lesbian & Gay Journalists Association (www.nlgja.org).

National Press Club (http://npc.press.org).

National Press Photographers Association (www.nppa.org/).

Native American Journalists Association (www.medill.nwu.edu/naja).

Online News Association (www.onlinenewsassociation.org).

Pew Center for Civic Journalism (www.pewcenter.org/).

Pew Internet and American Life Project (www.pewinternet.org/).

Poytner Institute for Media Studies (www.poynter.org).

Privacy Foundation (www.privacyfoundation.org).

Privacy International (www.privacyinternational.org/).

Project for Excellence in Journalism (www.journalism.org)

Radio and Television News Directors Association (www.rtnda.org).

Reporters Committee for the Freedom of the Press (www.rcfp.org).

Silha Center for the Study of Media Ethics and Law (www.silha.umn.edu/).

Society of Professional Journalists (www.spj.org).

Stanford Center for Internet and Society (http://cyberlaw.stanford.edu/).

U.S. Copyright Office (www.loc.gov/copyright/).

Index

Bold numbers, refer to figure.